A TOUGH APPRENTICESHIP:

SRI LNKA'S MILITARY AGAINST THE TAMIL MILITANTS 1979 - 1987

Channa Wickremesekera

A Tough Apprenticeship: Sri Lanka's Military
Against the Tamil Militants 1979 -1987

ISBN: 978 -0- 648-13490-9

Copyright © Channa Wickremesekera 2017
Typesetting by Brenda Van Niekerk
Cover design by Vernon Tissera
Published by the author
Cover image: Sri Lankan armed forces carrying out an
operation somewhere in the Vanni in the mid 1980s. Photo
credit: Raj Vijayasiri.

To
Kasy and Nalini

Table of Contents

Table of Contents

Acknowledgements

I am indebted to the generosity of Victor Melder who made available the resources of his library for my use. My thanks also go to Brigadier Bahar Morseth, Colonel Raj Vijayasiri, Colonel Kapila Ratnayake, Lieutenant Colonel Modestus Fernando and Lietenant Sudesh Ranawaka for sharing their memories with me. When I needed maps and plans drawn Rod Grigson readily obliged with his expertise and Raj Vijayasiri kindly supplied some valuable photographs. Last but not least, without BrendaVan Niekerk's type-setting skills and Vernon Tissera's cover design this book would not have been presentable.

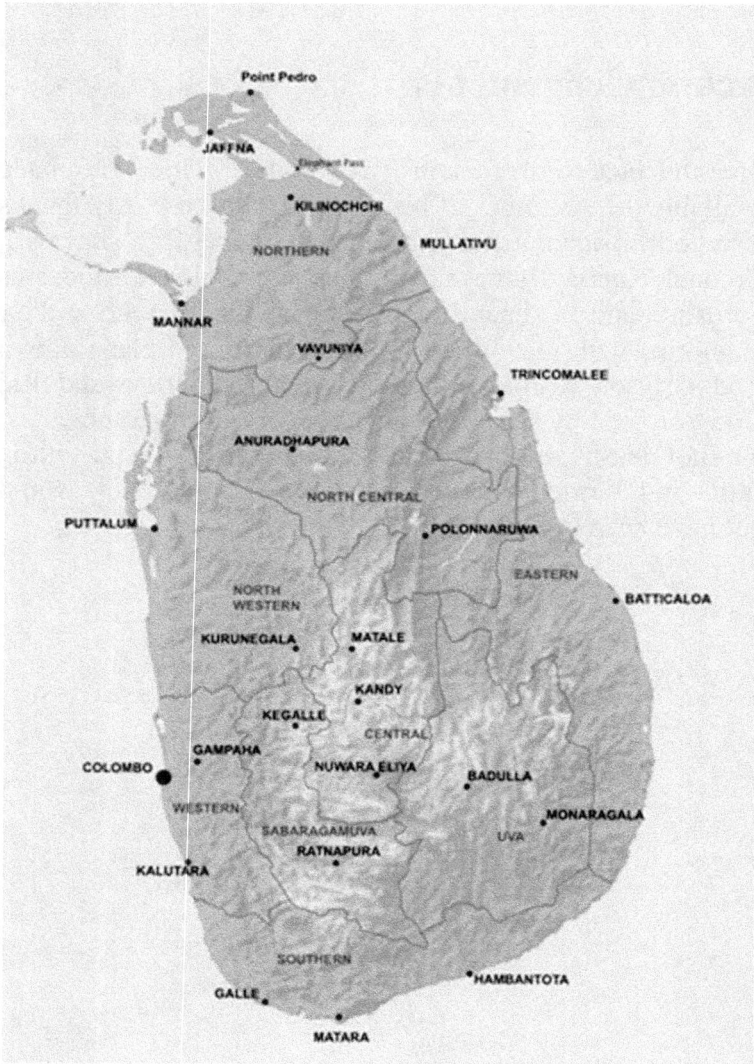

Map 1 Sri Lanka - Provinces and Major Cities

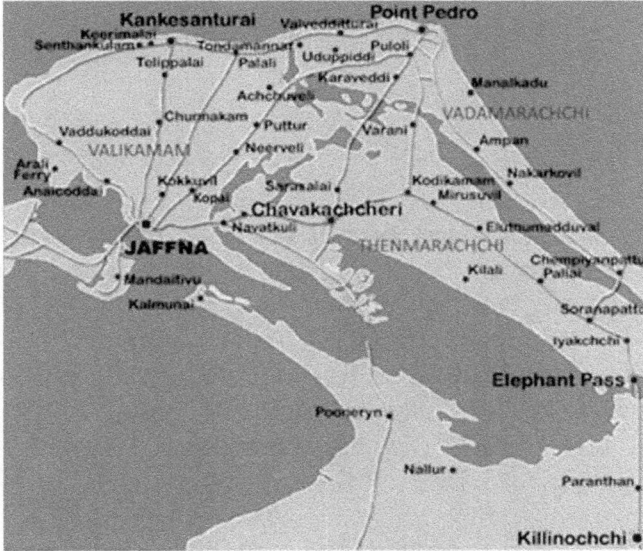

Map 2: Jaffna Peninsula

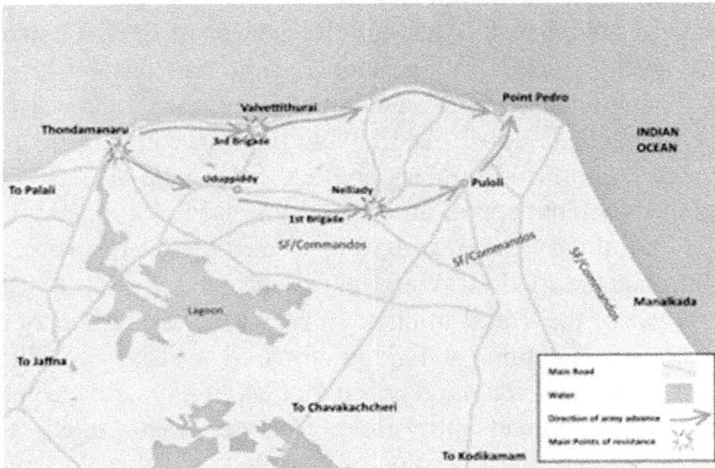

Map 3: 'Operation Liberation', May 1987

Map 3: 'Operation Liberation' May 1987

INTRODUCTION

In May 2009, the long separatist war in Sri Lanka came to an end. After months of gruelling campaigning, the Sri Lankan security forces finally cornered the military forces of the separatist Liberation Tigers of Tamil Ealam (LTTE) and annihilated them. All the leaders were killed while most of the hard-core cadres were either slain in battle or captured. The vast arsenal of the rebels was also captured or destroyed. As a military force, the LTTE was finished.

The army that defeated the Tigers in May 2009 was nearly thirty years in the making. Unlike many of her neighbours Sri Lanka did not possess military forces that were a continuation of organisations or traditions that predated Independence from European colonialism. Although the nucleus of the Sri Lankan army was formed by personnel from an establishment functioning under the British colonial government, this was not an active defence force. And having gained her Independence without a protracted military conflict, Sri Lanka also did not possess anything approximating to an 'army of liberation.' Even the military forces that were formed at Independence in 1948 remained modest in size, experience and vision. It is the escalation of the Tamil militancy that set the armed forces on the road to becoming a fighting force. The Tamil rebellion transformed the small, largely ceremonial, lightly equipped and inadequately trained army into a large, well-trained and equipped force that eventually defeated one of the most committed and dangerous guerrilla forces in the world. It was also accompanied by severe sacrifices. Nearly 30,000 members of the Sri Lankan armed forces lost their lives in the conflict while tens of thousands were wounded, many of them losing their limbs.

Coming to terms with an ethnic insurgency was a serious challenge to the fledgling Sri Lankan military. Much of the early phase of the conflict was spent in learning to negotiate the demands of combating urban terrorism and guerrilla warfare. Coupled with diplomatic and political challenges and constraints, it created fertile conditions for bloody setbacks and hard lessons. But it was also a period of growth. As the fighting intensified, the security forces expanded, acquiring more sophisticated weaponry and learning new skills. Operations became more complicated, their demands more taxing.

This book examines the conduct of military operations during the period spanning the origins of the Tamil insurgency and the premature termination of 'Operation Liberation' in June 1987. For the security forces this was a crucial period in the military conflict. It was the period in which they began to come to grips with the reality of a protracted ethnic insurgency suffering the vagaries of terrorism and guerrilla warfare. The period ends with the launching and the abrupt termination of Operation Liberation. The launching of that operation is a measure of the escalation of the military conflict as well as the evolution of the Sri Lankan security forces as a result of its demands. This was the first brigade -size operation launched by the Sri Lankan army which also involved the deployment of the bulk of the assets of the Sri Lankan navy and the air force in support of the ground operations. As such, it was an early and important watershed in the evolution of the Sri Lankan security forces and their struggle to meet the challenge posed by the Tamil rebellion.

The chapters are thematically and chronologically organised. They deal with the establishment of the Sri Lankan armed forces after Independence, the outbreak and escalation of the separatist movement, the response of the armed forces, the emergence of a stalemate in the north and the attempts to break the deadlock culminating with the launching of 'Operation Liberation'. They examine the challenges faced by the security forces and the measures taken to overcome them in the context of numerous constraints. The reprisals against civilians, the inevitable response of poorly trained and led troops are also examined. The conduct of Operation Liberation is scrutinised in some detail with an analysis of its successes and failures as well as its potential for success had the operation continued. The political and diplomatic context to the evolution of the conflict is also briefly discussed. The aim is to provide the reader with a multi-faceted view of the dynamics of the military operations during this crucial period.

It is hoped that this book will enrich the state of research into the military history of modern Sri Lanka. Despite its devastating impact on the country, the separatist war in Sri Lankan has generated little serious research. Only a handful of general studies have been produced and along with a few works with a more specific focus. In the first category are Paul Moorcraft's The Total Destruction of Tamil Tigers and Ahmed S. Hashim's When Counterinsurgency Wins: Sri Lanka's Defeat of the Tamil Tigers, while the latter is typified by the research of Raj Vijayasiri and Edward Amato. [1] Of lesser value are the works of Kamal Guneratne. C.A. Chandraprema and L.M.H. Mendis, which, while being informative, are marred by partisanship.[2] None of these works however, examine closely

---

[1] Raj Vijayasiri, 'A Critical Examination of the Sri Lankan Government's Counterinsurgency Campaign,(Master's Thesis, Fort Leavenworth, Kansas, 1999, Edward Amato, 'Tail of the Dragon: The Sri Lankan Efforts to Subdue the Liberation Tigers of Tamil Ealam', Master's Thesis, US Army Command and General Staff College, Fort Leavenworth, Kansas, 2002.

[2] Major-General Kamal Guneratne, *The Road to Nandikadal: The True Story of Defeating Tamil Tigers*, Colombo: Vijitha Yapa, 2016, C. A.

the early period of the military conflict with its significance for the development of the Sri Lankan security forces in to a fighting force. This period is mainly covered by a few articles, short studies and memoirs.[3] These offer valuable insights into this crucial yet little explored period in the conflict but they fall short of being sustained, systematic studies of the Sri Lankan armed forces' response to the emerging threat of Tamil separatism. This book hopes to make a positive contribution to filling this void.

The lack of current research makes the search for sources problematic; there is little guidance from previous studies as to the availability of sources. Apart from the few published narratives and studies, I have used a variety of published sources. Newspaper reports and articles are at the top of this list. Much of this stage of the conflict was reported more than

---

Chandraprema, *Gota's War: the Crushing of Tamil Tiger Terrorism in Sri Lanka*, Colombo: Ranjan Wijeratne Foundation, 2012, L. M. H. Mendis, *Assignment Peace: in the Name of the Motherland*, Nugegoda, Sri Lanka: Author publication, 2009.

[3] See for example Edgar O'Ballance, *The Cyanide War: Tamil Insurrection in Sri Lanka 1973-88*, London: Brassey's, 1990, which is an informative but somewhat rambling narrative of the early period of the military conflict. Rohan Gunaratna's *War and Peace in Sri Lanka* Colombo: Institute of Fundamental Studies, 1987, is also a useful though rather disjointed study. Thomas A. Marks carried out several, more focused studies of aspects of the conflict: These include 'Counter Insurgency in Sri Lanka: Asia's Dirty Little War', *Soldier of Fortune*, February 1987, pp.38-47, 'Insurgency and Counterinsurgency', *Issues and Studies*, August 1986, pp.63-104 and 'Sri Lanka's Special Forces', *Soldier of Fortune*, July 1988, pp.32-9. The memoirs and autobiographies include Cyril Ranatunge, *Adventurous Journey: From Peace to War, Insurgency to Terrorism*, Colombo: Vijitha Yapa, 2009, Gerry de Silva, *A Most Noble Profession: Memories that Linger*, Colombo: International Book House, 2011 and Tim Smith, *Reluctant Mercenary: the Recollections of a British ex-Army Pilot in the Anti-Terrorist War in Sri Lanka*, Sussex, UK: The Book Guild Limited, 2002.

studied and therefore, newspaper reports and articles are good sources of information, especially when they contain first hand observations, provided one is careful about the journalists' inclination to sensationalise and dramatise. The escalation of the conflict at a time without the Internet also spawned numerous publications which, despite the propaganda role of many of them, also function as valuable sources of information.

Last but not least are the interviews with some of the participants in the military operations. I have not conducted many of these mainly due to logistical issues and only some of these interviews have yielded useful information. Even thought the military conflict is over, many of the issues that begat that conflict still remain unresolved and people who speak of their roles in the war are extremely conscious of the impact their information may have on the continuing ethnic tensions. This is an issue that will continue to dog historians of the conflict for many years to come and this research has also been constrained by it.

Given the paucity of studies and our hazy knowledge of the sources, this work should be viewed more as an attempt to establish a base rather than conquer the territory. As such, it is to be treated as a pioneering work, not as a definitive history.

CHAPTER 1

A Modest Establishment: Sri Lankan Military Forces before the Tamil Insurgency

In January 1978 the Sri Lankan army held a 'Searchlight Tattoo' at the Sugathadasa Stadium in Colombo. The event was an opportunity to showcase the skills, expertise and hardware of the army. After hours of march-pasts by impeccably drilled soldiers and acrobatic displays by the more athletic men of the army, the massive crowd was treated to the *piece de resistance* of the evening: a mock assault on a terrorist stronghold. The facade of a fortress erected before a makeshift moat at one end of the stadium represented the terrorist camp which was to be taken by assault. The floodlights were dimmed while the troops prepared for the attack and when the lights brightened after a few minutes, the crowd gasped at the sight of two small artillery pieces set up on the lawn, surrounded by soldiers poised to open fire. Then the attack began. The guns opened up with dud shells, sending sparks streaming into the night air and making almost everyone in the audience clasp their ears in reaction. As the fire died down after a couple of minutes the shadowy figures of the commandos could be seen making the assault on the fort, entering it by means of cables strung across the moat. Then rifle fire erupted inside the fort as the brave men mopped up resistance. It was an exhilarating way to finish what has been a memorable evening.

If the tattoo was the army's way of showing the nation what skills and hardware it possessed to fulfil its role as the nation's defenders, the mock assault reflected the army's perception of the nature of the challenge it expected to face. The army was fighting insurgents, a home-grown enemy, not a foreign one. And the enemy was resisting in a conventional way by setting up fortifications complete with moat. It was to be reduced using the firepower of the big guns and the skill and daring of the commandos.

This perception had much to do with the circumstances of the establishment and evolution of the Sri Lankan army. Born in the flush of Independence in 1948 with no real challenge to stimulate its growth, it had evolved into a largely ceremonial outfit with very little combat experience.

1.1 A New Army for an Old Country

"Sinhalese are basically a peaceable lot with no great ability or desire for either work or war', Tim Smith, a British mercenary flying armed helicopters for the Sri Lankan air force in the 1980s, quipped sarcastically.[1] This is undoubtedly a sweeping generalisation, but the mercenary was just one in a long line of European observers who had spoken unkindly of the lack of martial qualities in the Sinhalese. "The Cingaleesch is a valiant foe over a fallen foe", a Dutch observer remarked contemptuously in the late 17th century. "[HE] would inflict on him even when dead another ten wounds, and be flaming after booty filling the air with shrieks and cries".[2] Barely a century later, a British observer was less scathing but hardly more charitable. Speaking of unsuccessful British attempts to raise a body of native soldiers in Sri Lanka in the late eighteenth and early nineteenth centuries, the Reverend James Cordiner

---

[1] Tim Smith, *Reluctant Mercenary: the Reflections of an Ex-Army Helicopter Pilot in the Anti-Terrorist War in Sri Lanka'*, Sussex, UK: The Book Guild Ltd, 2002, p. 107.
[2] Phillipus Baldaeus, *A True and Exact Description of the Great Island of Ceylon*, a New and Unabridged Translation from the edition of 1672, *Ceylon Historical Journal* vol. 8, July 1959 - April 1959, p. 236.

explained that the project failed because 'a life of military discipline proved, in the highest degree, irksome and uncongenial to their habits'. [3]

Such observations must rankle the present day Sinhalese, especially in the aftermath of the comprehensive victory over the Tamil Tigers, once considered one of the most dangerous guerrilla forces in the world, by armed forces which were almost exclusively Sinhalese. Those who are familiar with Sri Lanka's struggle with colonial powers will find such observations uncharitable to the armies of Sinhalese kings who had given the military forces of three European powers a hard time and, at times, sound thrashings. But such indignation, while perfectly justified, also miss the point. The Europeans' perceptions were informed by a particular conception of military competence obtaining in contemporary European military culture.

All the above observers saw soldiering as a profession, something one was trained to follow in body and mind according to certain expectations: regimented life, steadiness under fire and obedience to orders, all hallmarks of the European military system that had evolved since the 16th century. This is what, in their opinion, the Sinhalese lacked and what made them poor soldiers. Few of the Sinhalese armies or soldiers these observers had witnessed could dispel this prejudice. The army that crushed the Tigers was only beginning to evolve in the 1980s and before that there existed no military tradition to satisfy western martial expectations.

This is not because the Sinhalese are congenital pacifists. Sri Lanka's history is replete with bloody conflicts, murders and

---

[3] James Cordiner, *A Description of Ceylon, an Account of the Country, Inhabitants and Natural Productions*, 2 vols., New Delhi: Navrang 1983, First Published by Longman, Hurst, Rees and Orme, Aberdeen, 1807, v.1, p. 92.

downright butchery. However, in Sri Lanka, wars were fought not by professional soldiers but by peasants who left their fields to fight for their kings and returned to their fields when the campaign was over or when their food supplies ran out. The only professional fighters were foreign mercenaries who functioned more as a palace guard and an elite body than as an army. There were no castes or other social groups who had a martial "calling". The largely subsistence economy and military organisation based on feudal obligation did not permit wars that lasted very long. It was neither profitable nor practical to keep the peasants off their land for too long. This in turn afforded little scope for martial traditions to develop among the peasantry through long association with fighting while also discouraging the evolution of a social organisation that accommodated a warrior class or caste. Finally, the role of Buddhism should not be discarded. Buddhism emphasised non-violence and detachment. Although this has never stopped Sri Lankans from killing each other it has arguably also prevented war from becoming a focus of their culture.

Even the advent of the European powers in the 16th century did not change this. The sixteenth and seventeenth centuries were crowded with bloody confrontations between European and indigenous armies, leading to the demise of tens of thousands of lives and leaving swathes of the island devastated and depopulated. The wars with the Portuguese in particular exposed the population in the southwest and the central highlands of the island to intense fighting on an unprecedented scale and intensity. But this doesn't seem to have led to the militarisation of a segment of the society or any appreciable change to the structure of the state that governed that society. The peasants, taken from the fields in times of war, returned to the fields at the end of the campaign, toughened by the experience but without acquiring the outlook of fighters while the system of patronage and service continued. A few decades of widespread bloodletting were not sufficient to change centuries of nurturing.

Due to their reputation for lack of martial qualities, the British, when they mastered the entire island of Ceylon (as Sri Lanka was then called by the British) did not look favourably upon the 'natives' as potential recruits for their military establishment in the country. The choice fell on the Malays who were considered more energetic and aggressive. After the conquest of the island in early nineteenth century, the British maintained a 'native' military establishment, the Ceylon Rifle Regiment, composed largely of Malay soldiers. The unit was disbanded in 1874 but seven years later the British raised a new unit called the Ceylon Light Infantry Volunteers (CLIV) which incorporated Europeans and Ceylonese within its ranks. In 1910 the CLIV was renamed the Ceylon Defence Force (CDF). During World War II the CDF performed guard duties in the city of Colombo and manned the artillery at the harbour. Despite having a Ceylonese component in the CLIV and the CDF the British ensured that their officer cadre were exclusively British. The natives, even the most westernised, had no place as military leaders under their colonial masters.[4]

After Independence in 1948, the new government of Ceylon had the task of raising an army for its defence. In their enthusiasm for building a new army for a new nation some of the military planners in Colombo envisaged a fully fledged modern army with all the trappings, but economic and political realities militated against that. Ceylon simply did not have the funds to build a modern army. There was also no great need for one; there was no conceivable enemy in the region and even if there was one, a defence agreement with Britain ensured that

---

[4] Donald L. Horowitz, *Coup Theories and Officers' Motives: Sri Lanka in Comparative Perspective*, Princeton, NJ: Princeton University Press, 1980, p. 63.

the island's former colonial masters would come to its aid in the event of an emergency.[5]

So the new nation had to do with an army that was of modest proportions. Naturally the CDF came to form the nucleus of the new army. The first units to be raised were an artillery regiment, an infantry regiment (the Ceylon Light Infantry or CLI), a signals squadron, two engineering units, a service unit, a military police unit and a recruiting unit. The British, although they had given up the reigns as the rulers, still had a measure of influence in the running of post-independence Ceylon. The army was raised under close advice from the British military. A British officer was even appointed the first Commander-in-Chief. The soldiers were all equipped with discarded British hardware; the infantry carried .303 Lee Enfield rifles and a few Bren guns while the artillery was equipped with a handful of 4.2 inch mortars and 3.6 inch anti-aircraft guns.[6]

1.2 The Army Evolves

Gradually the army evolved. In 1956 a new infantry unit, the Sinha Regiment was formed. Another regiment, the Gemunu Watch, was added in 1962. An 'armoured squadron' was raised in 1955 and elevated to the status of a regiment, a rather grandiose title for a unit whose 'armour' consisted of several Ferret scout cars and Daimler armoured cars.[7]. But soon a measure of greater respectability was added by the acquisition of a number of Saracen armoured personnel carriers that the

---

[5] Brian Blodgett, *Sri Lanka's Military: the Search for a Mission*, San Diego, California: Aventine Press, 2004 p. 25.

[6] Ibid, pp. 26-8.

7 '1st Reconnaissance Regiment Sri Lanka Armoured Corps', *Army Magazine*, 30th Anniversary Issue October 10, 1979, p. 23, Jagath P. Senaratne, Sri Lanka Armoured Corps: *60 Years of History 1955 - 2015, Colombo; Sri Lanka Armoured Corps*, 2015, pp. 9-10.

British army was happy to relieve themselves of as they withdrew from Singapore and Malaya[8].

Thus for the first time, an independent Sri Lanka had a standing army, structured, trained and equipped after the Western fashion. It was created virtually out of nothing. The CDF that was there before - and which provided many of the cadres for the new army- was hardly a regular fighting force. It saw little action and was more a body that provided guards than soldiers. The new army created in 1949 was to be a real, standing army trained and organised like any modern army, something Sri Lanka had never had.

The size and the armaments of the new army reflected its role in independent Sri Lanka. Armed mainly with WW II rifles, Bren guns, artillery, and scout cars, the few hundred soldiers were more a ceremonial force than an effective defence force. Even the slight increase in numbers as a result of new regiments did not make it a force that could intimidate any potential invader. Accordingly, the doctrine of the army shifted from preparing for a conventional confrontation to maintaining internal security. There was also focus on preparing for a long guerrilla campaign, something made fashionable by the eruption of guerrilla wars in many parts of the former European colonies in the 1950s and '60s. In the event of an invasion - which the military was not equipped to repel - the army was to retreat into the jungles and from there wage an irregular war to cripple the enemy.[9]

No invasion materialised but there were a few internal disturbances in the 1950s and the '60s to keep the little army busy. None of these amounted to a serious armed threat and they were handled without any great difficulty. Thus, during the

---

[8] *SL Army 50 years on*, Colombo: Sri Lanka Army, 1999, p. 92.
[9] Ibid., p. 167.

Hartal in 1953 and a wave of strikes and ethnic unrest in the late 1950s the army "aided the civil power" in facing the challenge of maintaining law and order as well as running essential services smoothly. In 1961 the military was also called in to keep order when the Federal Party launched a *Satyagraha* in the North and the East. In the 1950s and '60s the army was also stationed in the north to thwart illegal Indian migrants from entering Sri Lanka.[10]

These operations required minimum use of force against largely unarmed 'adversaries.' One exception to this was the operations during communal riots in 1958. The military was compelled to open fire on several occasions on armed rioters causing considerable casualties. In one particularly serious incident an army unit opened fire at a 600 - strong force of rioters approaching the town of Anuradhapura with the aim of killing Tamils and destroying their property, killing and wounding several miscreants.[11] This however, was an aberration. The military was usually able to intimidate protesters and demonstrators by their mere presence.

Until the early 1970s then, the Ceylon Army – as it was known at the time - enjoyed the status of a largely ceremonial force that marched and presented arms during parades and helped the police to intimidate strikers and demonstrators and occasionally rioters. The men were generally armed with the antiquated .303 rifle with Light Machine guns at company level. Dressed in 'web belt, ankles and khaki uniform and the ill-fitting steel helmet' they were also burdened with a heavy pack of over 30kg.[12] When not parading and maintaining public order on the

---

[10] Blodgett, *Sri Lanka's Military*, p. 34.

[11] Tarzie Vittachi, *Emergency '58: The Story of the Ceylon Race Riots*, London: Andre Deutsch, 1958, p. 67, T. D. S. A. Dissanayake, *War or Peace in Sri Lanka*, Colombo: Popular Prakashan, 1995, vol 2, pp. 258-9.

[12] Cyril Ranatunge, *Adventurous Journey: from Peace to War, Insurgency to Terrorism*, Colombo:Vijitha Yapa, 2009, p. 40.

odd occasion, the soldiers spent much of their time playing sports and engaged in social activities. The last thing on their minds was fighting.

However, even though the army's fighting potential hardly changed until the early 1970s there were other changes that had portends for the future. This was the steady Sinhalisation of its ranks. The change in composition reflected political events. Since the mid-1950s, communalism was beginning to play an important role in the political discourse of the island. The leaders of independent Sri Lanka had come to realise that in an electoral system depending on majority votes, communalism could be a potent tool in mobilising people. The majority of the people being Sinhala the state began to reflect the interests and sensibilities of the Sinhala people. The nationalist MEP government that came to power in 1956 abrogated the defence pact with Britain and went on to implemented policies that consolidated the power of the majority community. The Language Act of 1956 which made Sinhala, the language of the majority of the people the official language, was a first step in the direction of Sinhalisation of the state which ignored the multi-ethnic nature of the society that had gained independence from the British. And as the state institutions Sinhalised, so did the Army which began to gradually acquire the role of the armed forces of a Sinhala state, the Sinhalisation of the forces simply completing the organic link between the two. The percentage of Sinhalese Commissioned Officers in the Ceylon Light Infantry rose from 5% in 1949-51 to 96% in 1963-69 while the number of Tamils dropped from 18% to 12%. The percentage of Buddhists, who were almost exclusively Sinhalese, rose from 34% in 1949-51 to 89% in 1963-69, while Christians dropped from 59% to 7%. Somewhat similar trends were seen in the Artillery as well[13]. After the few Tamil officers retired few if any joined the army and the enlistment of Tamils

---

[13] Horowitz, *Coup Theories and Officers' Motives,* pp. 69-70

in the ranks was reduced to a trickle. By the late 1970s the army was almost exclusively Sinhalese.

There were other aspects that reflected and aided this process. The names of the new military units formed after 1956 – Sinha Regiment and Gemunu watch – had a clear Sinhala nationalist flavour. The mascot of the CLI was the elephant Kandula named after the famous war elephant of king Dutugemunu who is credited with unifying the country by defeating the Tamil king Elara[14]. After 1956 the words of command were also given in Sinhalese[15]. The growing parochialism was further aided by the discontinuation of sending officer cadets abroad for training. After Independence and even after the termination of the defence pact with Britain, selected officer cadets were sent for training at Sandhurst but this was discontinued in the late 1960s in favour of local training. The rising economic cost was the main reason behind this decision but its impact was far reaching. Not only did it deny the cadets an opportunity to train at the highest levels but it also robbed them of the chance to experience and absorb the military culture of more professional military establishments.

The army then, was gradually becoming a Sinhala army in composition and culture. Its symbolism was borrowed from a Sinhala universe and its identity was derived from Sinhala ethnicity. Army publications were also usually in English and Sinhalese. By the time of the outbreak of the war with the Tamil militants the military was almost exclusively a Sinhalese force with a Sinhalese outlook.

The increasing Sinhalisation and the growing communal tensions also served to redefine the military's role. In the late

---

[14] This was the practice since 1961. Malaka Rodrigo, 'Kandula - the Little Elephant of the Army'', *Sunday Times*, 21. 06. 09, http://www.sundaytimes.lk/090621/Plus/sundaytimesplus_13.html.
[15] *Sri Lanka Army*, p. 97.

1950s there were only two small military detachments stationed in the north, one at Palali and one in Mannar. The military presence in the north had come as a result of the Sinhala - Tamil riots in 1958. This deployment may not have been communally motivated as other units were placed in the western, southern and eastern parts of the island to quell civil unrest.[16] However, there were signs that military deployment in the North was beginning to acquire greater importance in the early 1960s. In 1963, the Defence Secretary N. Q. Dias reportedly wished to establish a greater military presence in the north by setting up of a ring of army camps, "all the way from Arippu, Marichchikatti, Pallai, and Thalvapadu in the Mannar District, through Pooneryn, Karainagar, Palali, Point Pedro and Elephant Pass in the Jaffna District, and on to Mullaitivu in the Vavuniya District and Trincomalee in the East". He also wanted to set up a task force ostensibly to watch smugglers and illicit immigration but in reality to watch the Tamils in the north and prevent them from challenging the government's authority.[17] This was a reaction to the Satyagraha agitation in 1961 where the Tamils protested vehemently against the government's Sinhala only policy. The agitation was seen as a sign of things to come, an extraordinarily prescient view. Dias' proposal for the ring of camps was not implemented but a task force was set up to curb illegal immigration, bringing a greater military presence to the north. The proposals show the emerging conception of the army as an instrument to cow the Tamils, at least in some powerful sections of the government of the time. This would gain greater currency with the outbreak of hostilities with the Tamil militants.

---

[16] Anton Muthukumaru, *The Military History of Ceylon – an Outline*, New Delhi: Navrang, 1987, p. 165 and pp. 170-2.

[17] Neville Jayaweera, 'In to the Turbulence of Jaffna: a Chapter extracted from the Author's unpublished Memoirs titled 'Dilemmas'', *The Island*, 5. 10. 08, http://www.island.lk/2008/10/05/features2.html

1.3 First Blood

Ironically, when the army was finally forced to fight it was not against Tamils but against Sinhala insurgents. In April 1971 the left-wing JVP launched an island-wide armed assault on the government. It placed the military under severe strain. The army was not facing unarmed demonstrators or immigrants anymore; the enemy was a highly motivated group of young men and women, armed with somewhat primitive - yet deadly - weapons. And they were attempting a takeover of the government by launching simultaneous attacks on police stations across the island. For a while the army was stretched, lack of experience in handling an insurgency and the inadequacy of transport critically handicapping the soldiers. Still, after the initial shock the army was able to wrest the initiative from the insurgents. The JVP cadres, despite their high level of motivation, were poorly armed and trained. The army's rifles and Bren guns were more than a match for the ill-aimed shot guns and home-made hand bombs of the enemy.[18] But the victory was not bought without cost. A number of officers and men were killed and wounded before the militants could be neutralised.[19]

The JVP insurrection was a milestone for the army for a number of reasons. It provided the troops with real operational experience for the first time, no matter how fleeting the fighting may have been. It was also a serious challenge. The military establishment had to co-ordinate an island wide deployment in

---

[18] The Army's firepower had been boosted by the purchase of 990 semi-automatic rifles from Australia. These are the SLRs seen in photographs of Army operations during the JVP insurrection. Blodgett, *Sri Lanka's Military*, p. 53. For a brief but excellent analysis of the disparity between the insurgents and the army see Senaratne, *Sri Lanka Armoured Corps*, pp.97-99.
[19] See A. C. Alles, *The J.V.P. 1969-1989*, Colombo: Lake House, 1990, James Manor and Gerald Segal, 'Causes of Conflict: Sri Lanka and Indian Ocean Strategy', *Asian Survey*, 25, no.12 (December 1985) pp. 1165-1185.

co-operation with the other arms of the government while the soldiers faced a real combat situation for the first time, providing valuable exposure for an army that was only used to marching and presenting arms. The success of the troops, however, owed more to its superior firepower and the amateurish strategic and tactical approach of the insurgents rather than due to any exceptional professionalism on the part of the army. In a clash between amateurs those with the better guns won. Still, the experience was valuable.

However, the biggest consequence of the insurrection was the strengthening of the security forces. Many countries felt sympathetic towards Sri Lanka in her hour of need and demonstrated this by donating weapons and equipment, boosting the defensive and offensive capabilities of the army. During 1971 and 1972 Sri Lanka received over 30,000 rifles and semi automatic rifles from China while Australia also chipped in with 5000 rifles. Russia donated 12 light mortars, Yugoslavia sent four pieces of 76 mm mountain guns and China provided 30 pieces of 85-mm artillery. In 1972 the USSR also donated ten BTR152 APCs while Sri Lanka purchased 18 ferret scout cars from the UK to boost the existing fleet of Saracens and Ferrets. The UK also supplied several Saladin armoured cars.[20]

When the Tamil insurgency raised its head the army was much better armed than it had ever been. But it was still small with no more than 9,000 troops. And despite the upgrading following the JVP insurgency, it was still woefully backward in comparison to any modern army in terms of training and equipment. Although the donations and acquisitions in the wake of the '71 insurgency had boosted the army's firepower, the arms and equipment the army received was what the countries that donated them thought Sri Lanka needed, or what

---

[20] Blodgett, *Sri Lanka's Military*, pp. 68-9.

they could conveniently dispose of, not necessarily what the army required. The armoured cars and the BTR 152s were not in production anymore while the 85mm artillery pieces were of dubious usefulness in guerrilla warfare. Still, some of the equipment like the light mortars, the mountain guns and the armoured personnel carriers did have a counter-guerrilla or counter-insurgency function. But generally, the equipment reflected the convenience of the donors rather than the needs of the recipients.

Moreover, these changes were more cosmetic than fundamental. In many ways, the army progressed very little. It continued to recruit heavily from the Sinhalese community and struggled with improving operational and organisational aspects. A gradual rearming of the infantry was taking place, but there was no uniformity, some units being armed with the Chinese T-56 while others received SLRs received from Australia while still others remained armed with the vintage .303s. The artillery remained the same while armour improved little beyond receiving a consignment of reconditioned armoured and scout cars in the late 1970s.[21] In his report to the Ministry of Defence in 1978, the army commander bitterly complained about the training standards of the army. Although the individual training of the men was satisfactory collective training lagged behind. He also pointed out that more than 30% of the vehicular strength of the army was more than fifteen years old and that "specialised military equipment in specialised units such as Armour, Artillery and Engineers is outdated and in most instances over 20 years of age'[22]. To make matters worse, the army had carried out no weapons training between 1972 and 1978 largely due to a lack of ammunition and the artillery regiment practised indirect firing

---

[21] *Sri Lanka Army*, p. 94.
[22] Ibid., p. 286.

for the first time in October 1978.[23] No major organisational or structural changes took place other than the division of the army into six area commands. Even this was done in a manner that ensured chaos rather than order in the case of an emergency as not all the units under an area command were stationed in that area. It demonstrated a lackadaisical approach which smacked of complacence and a lack of vision.

The only innovative measure to come out was the formation of a commando unit in 1977. At first, this "regiment" consisted of only a handful of men seconded from various units, the best men from each unit. Trained by former members of the elite SAS, the commandos were divided into two groups of 20 men and officers each. One was an assault group, the other a sniper team[24]. This however had little to do with preparation for the looming confrontation with the Tamil militants but was initiated as a safeguard for the tourism industry. The main focus of the commandos was the protection of the Katunayake International Airport. They were more rigorously trained than the regular soldiers but were mostly armed with only T56 assault rifles. [25]

If the army was indifferently prepared for war, the air force and the navy were in a worse state. When the JVP launched its uprising in 1971 the Sri Lankan "air force" consisted of a fleet of aging fixed wings aircraft and four helicopters. The fixed wing aircraft consisted of several De Haviland Herons and Doves which were transport aircraft, about a dozen chipmunk trainers and jet provosts while the helicopters were three Bell Jet rangers and one Hiller helicopter[26]. During the insurgency the air force brought the jet provosts out of storage and armed

---

[23] Blodgett, *Sri Lanka's Military*, p. 71.
[24] Shamindra Ferdinando, 'Armless Veteran speaks of War and Peace', *The Island*, 27. 10. 02, http://www.island.lk/2002/10/27/featur08.html
[25] Ibid.
[26] Blodgett, *Sri Lanka's Military*, pp. 56-7.

them with Browning machine guns and rockets while the helicopters were provided with Bren guns to act as improvised gunships.[27] Although very rudimentary, these measures enabled the government forces to maintain their superiority against a lightly armed insurgent group. Like the army, the air force too benefitted from the generosity of foreign countries in the wake of the JVP uprising. The Russians provided six MiG-15s, one MiG-17 and two Ka-26 rescue helicopters. The British sent six Bell 47 Gs armed with 7.62mm machine guns which were immediately used against the rebels. [28] But after 1971 the air force joined the other forces in languishing in inactivity. The Bell 47Gs went into storage and the Bell Jet rangers were used to provide "helitours" scenic flights for tourists.

Although Sri Lanka's first line of defence from external threats was her shore, her naval defences were almost non-existent. Up to the insurrection in 1971 the 'navy' possessed only two aging frigates and about two dozen patrol crafts. Following the insurrection Sri Lanka received two Shanghai class patrol craft from China. Three more were purchased the following year. Several patrol boats were also purchased from the Soviet Union.[29] Expansion beyond this was not seen necessary and financially unaffordable.

This was the military establishment that had to take on the Tamil insurgency. They were of meagre proportion and limited experience which made for modest expectations. By the late 1970s the Sri Lankan military forces had come to perceive their main combat role as one of defending the state from internal threats. There was no conceivable external enemy and even if there was one, the forces at the disposal of the state were in no state to fight them off. Past experience also suggested that any

---

[27] Ibid., pp.72-4, Peter Steinemann, 'The Sri Lanka Air Force', *Asian Defence Journal*, Feb. 1993, pp. 54-56,
[28] Blodgett, *Sri Lanka's Military*, pp. 72-6.
[29] Ibid., pp. 79-80.

future threat would be internal. And the JVP insurgency had set the tone for future expectations. The JVP had attacked police stations in strength and when these failed had taken refuge in the jungle. There was no protracted guerrilla campaign that bled the state dry. Fighting was fairly straightforward; the JVP gave the security forces something to aim at or some place to look for them. The moated fort at the military tattoo was more than mere theatre. It was also expression of the military's vision of future combat.

But the future has a habit of springing unpleasant surprises. When the Tamil militancy broke out, it was to be a very different challenge to the one posed by the JVP insurgents.

CHAPTER 2

A Ragged Rebellion Comes of Age

In hindsight, N.Q. Dias appears to have been remarkably prescient about unrest in the North becoming the main threat to national security in the coming decades. While the soldiers were firing their dud rounds at an imaginary enemy at the Sugathadasa Stadium a real insurgency was already stirring in the predominantly Tamil north. In the years to come, this would grow into proportions that would beggar the scariest nightmares of N. Q. Dias. But this was neither inevitable nor unavoidable. The problem with Dias' perception is that at its foundation was the assumption that the Tamils were a people to be feared, and therefore, best kept in check. Had the Sri Lankan state, of which Dias was an important official, perceived the Tamils as a people who also had fears, events may have taken a happier turn for Sri Lanka.

2.1 The Backdrop to Rebellion

When Sri Lanka gained Independence from Britain in February 1948 there were two clear cultural, linguistic and ethnic zones in the island. The northern and Eastern provinces were predominantly Tamil- speaking with the majority of the population ethnic Tamils, while the majority of the inhabitants of the rest of the island were ethnic Sinhalese who spoke Sinhala. In the central highlands, there was a large concentration of Tamil speaking people working in the tea plantations. The Eastern Province was also home to large Sinhalese and Muslim populations, the latter using Tamil as their first language. The majority of the Sinhalese were

Buddhists while the Tamils were mainly Hindus with both communities having large minorities of Christians.

Of the Tamils, those in the central highlands were of recent origins, their ancestors brought to Sri Lanka as indentured labourers by the British, but the Tamil population in the North and the East had a longer history reaching back to times of political realties very different to those which existed in late 20th century. It is fair to say that both Tamils and Sinhalese had arrived in the island from the neighbouring subcontinent at a time when little more than the shallow stretch of ocean separating the island from the subcontinent stood in their way. The first great kingdom on the island, Anuradhapura, was associated with the Sinhalese people even though it is conceivable that there would have also been a considerable Tamil presence on the island as a consequence of trading activities as well as frequent hostile incursions by adventurers. However, the domination of the north by Tamil speaking people dates from the establishment of Tamil rule in the north as a result of the expansion of the Chola Empire of South India in to the island in the 10th century, leading to the establishment of a Tamil kingdom in the north. As Sinhalese rule retreated southward and finally into the central highlands, Tamil influence also spread along the eastern coast laying the foundation for a Tamil speaking north-eastern region on the island. Apart from a short-lived conquest by the Sinhalese in the 15th century, the northern Tamil kingdom based in Jaffna maintained its independence from the Sinhalese until its conquest by the Portuguese in 1619. From the Portuguese the kingdom passed onto the Dutch in 1658 and then from the Dutch to the British in 1796, ensuring its political independence from Sinhalese rule. The Eastern seaboard, on the contrary, remained under the orbit of Sinhala rule in the central highlands but culturally and linguistically, it had developed in to an extension of the Tamil dominated north.

The British brought the entire island under their control by 1818 but this did little to bring the Tamil and Sinhalese communities together. At the elite level Tamils and Sinhalese fraternised, thanks to English education and western culture but beyond that there was little interaction between the two communities. Leading up to Independence, Tamil and Sinhalese elites became increasingly aware of the electoral strengths and weaknesses of their communities in an independent Ceylon where political power was to be decided by the ballot box. Predictably, after independence from Britain in 1948, the Sinhalese and Tamil leaders devoted their energies to harnessing the electoral strengths of their communities to ensure and further their political careers rather than building a nation all communities could feel secure in. The result was a government dominated by and reflecting the wishes of the Sinhalese people as the majority community. In the Tamil speaking areas, particularly in the north, Tamil political parties opposed to majority domination gained in influence.

Events took a turn for the worse with a change of government in 1956. The new Prime Minister Solomon Bandaranaike delivered on his election promise to make Sinhala the official language. Bandaranaike, like Sir John Kotelawela he replaced, was no racist, but the demons released by his opportunism were too difficult to control. The passing of the Official Language Act No. 33 resulted in a ridiculous situation that forced Tamils to learn Sinhalese in order to get jobs in the government. Predictably, the Tamils were outraged. They saw it, quite understandably, as an attempt to marginalise them in the new nation. This resentment was only exacerbated by the failure to negotiate a settlement with the government. Attempts by Sri Lankan governments to redress Tamil grievances through a measure of power sharing in 1958 and 1965 were stymied by the increasingly powerful lobby of Sinhalese nationalists led by Buddhist monks. This stratum of Sinhalese society had emerged as a major force during the agitations in favour of

Sinhala as the official language. Now they asserted their power by forcing the government to withdraw the concessions.[1] This made many Tamils feel that the Sri Lankan government had little interest in going against the strident voices of the Sinhalese nationalists.

Their initial protests were peaceful, little more than the usual mass demonstrations and sit-ins. They tried it in 1958 and 1961, the latter an exceptionally successful *satyagraha* which attracted widespread support from the Tamils in Jaffna as well as Batticaloa in the Eastern province. The Sri Lankan government's response, however, was stern. The protesters were treated like contemptible upstarts. And to remind them of their place the military was called in, and the Satyagrahis were handled with harshness, sometimes even with brutality.[2] The resistance gradually petered out. The sit-ins had galvanised the people in the north into action but achieved little more than ensuring that the government remained firm in its stance.

By the early 1970s the increasing effeteness of the tactics of the Old Guard of Tamil political leadership was beginning to frustrate the youth. The old leaders were too steeped in traditional methods of persuasion. Sit-ins and petitions did not work with a government that seemed too closely bound to the majority community. They needed something more imaginative - and powerful.

2.2 The Rise of Tamil Militancy

The discontented Tamil youth gradually became convinced that the Sinhalese state had to be fought with fire, not with resolutions and speeches. In 1971, under the leadership of the

---

[1] David Little, *The Invention of Enmity*, Washington: D.C.: U. S. Institute of Peace, 1994, pp. 68-74.
[2] S. Ponniah, *Satyagraha: The Freedom Movement of the Tamils in Ceylon*, Jaffna: A. Kandiah, 1963, pp. 54-94.

JVP, the youth in the predominantly Sinhalese South had risen against the state trying to overthrow the government in a violent uprising. It has been crushed with greater violence but the abortive uprising provided an inspiration to many militant Tamils. The Sinhalese state, they were beginning to believe, had to be jolted into its senses by violence. Buoyed by this conviction, the expressions of Tamil discontent now began to acquire a more aggressive form. Militants hoisted black flags, torched government buses and property, felled trees across roads and even occasionally lobbed a bomb or two made of chemicals stolen from schools and colleges.[3] The days of expressing discontent by sitting down on the road were over.

The growing militancy was led by a number of youth organisations that sprouted in the late 1960s and early 1970s. One of the most prominent was the Tamil Youth Front which was seen as an unofficial youth wing of the main Tamil political party the Tamil United Front. However, at this stage, only a few of these groups embraced violence as the main weapon in their struggle. The Liberation Tigers of Tamil Ealam or LTTE which was to gain notoriety later as a ruthless and powerful Tamil guerrilla group was one of them. It began its life as the Tamil New Tigers (TNT) formed by Velupillai Prabhakaran a youth from Velvetithurai in Jaffna. The TNT and another group under a youth named Nadarajah Thangathurai were the two leading militant groups at the time. Both were committed to armed struggle as a means of achieving Tamil independence.

The early acts of violence by the militants, though marking a decidedly aggressive turn in Tamil dissent and protest, did not lead to bloodshed or loss of life. That came in 1975 with a murder that underlined the militants' growing dissatisfaction

---

[3] M. R. Narayan Swamy, *Tigers of Lanka: from Boys to Guerrillas*, Delhi: Konark Publishers, 1994, p. 26.

with the Tamil leadership. The victim was the Jaffna Mayor Alfred Duraiappah, a popular figure in Jaffna. A member of the ruling Sri Lanka Freedom Party (SLFP) he had built for himself a secure and firm base of support in the prestigious electorate of Jaffna through his brand of populist politics and the patronage he received from the government. [4] However, his contributions to the community were not enough to spare him the wrath of the militants. To them, Duraiappah was a government stooge, a sell-out who had betrayed his people for the crumbs from the government's table. Some even held him responsible for ordering a police crackdown on a Tamil conference in 1974 which had led to the death of seven people. In the eyes of the budding militants especially those now coalescing around Prabhakaran, he was as good a traitor as any to be made an example of. Accordingly, on 27th July, 1975, three young men shot him dead when he arrived at the Kovil in Ponnalai in Jaffna for his customary Sunday prayer signalling the beginning of a new, violent phase of the conflict.

The murder of Duraiappah was a huge shock to the government. The government had been aware of the extremist cells in Jaffna for some time. However, they did not expect their extremism to take so bloody a turn, claiming such a high profile victim. And to make matters worse, Duraiappah's murder was followed by more killings. On 2nd July 1976 the LTTE gunned down N. Nadarajah, a suspected police informant. They followed it up by shooting police constable Karunanidhi dead at Maviddapuram in Jaffna in February 1977. The two constables appointed to investigate Karunanidhi's murder met a similar fate. On May 18th the same year they were shot dead by several LTTE gunmen who came

---

[4] 'The Murder of Alfred Duraiappah', University Teachers for Human Rights (Jaffna),
http://www.uthr.org/Book/CHA02.htm#_Toc527947383

on bicycles.[5] These events were further compounded by the adoption of a more aggressive approach by the mainstream Tamil politicians. The main Tamil political Party, the Tamil United Front (TUF) changed its name to Tamil United Liberation Front (TULF) and gave vent to the growing militancy among the Tamils by passing a resolution at its congress in Vaddukoddai in May 1976 calling for a separate state of Tamil Ealam as the solution to the Tamils' marginalisation in Sri Lanka.[6] Not only were the militants spilling blood but their demands were becoming more extreme and threatening and influencing the politics of mainstream Tamil parties. Matters became further complicated when the Tamils voted overwhelmingly for the TULF's platform of separation at the General Election in July 1977. The TULF won 18 out of the 24 seats they contested in the north and the East, including all the seats in the Northern Province, the Tamil heartland. Separation was not something that only the militants and the politicians wanted anymore. It now had the endorsement of the Tamil people as well.

These were worrying signs for the Colombo government, now under a new Prime Minister, J. R. Jayawardena. The task of nipping the incipient armed rebellion in the bud fell to Inspector Bastiampillai, a police officer with a fierce reputation. But when Bastiampillai was gunned down along with his escort in the jungles in the northwest of Sri Lanka where he had ventured in search of the militants, the worry turned to alarm. It approached panic proportions as the militants continued their violent spree in the following months. In May 1978 they killed Inspector Pathmanathan at his residence in Jaffna and in June that year retired Inspector Kumaru was shot dead. On September 7th the militants changed tactics. The AVRO 748 of

---

[5] Narayan Swamy, *Tigers of Lanka*, p. 58.
[6] T. D. S. A. Dissanayake, *War or Peace in Sri Lanka*, Colombo: Popular Prakashan, 1995, vol. 2, p. 268.

the domestic airline Air Ceylon was blasted on the tarmac of the Ratmalana airport using a time bomb. The original plan seems to have been to destroy the plane on its way to the Maldives but a delay in catering had saved a bigger disaster. The culprits were never caught and the incident sent a strong message to the authorities that the militants were becoming increasingly bolder and deadlier.

Not content with assassinations and blowing up planes, the militants also turned their attention to robbing banks. The reason was simple. They needed funds to sustain their movement. Besides, it was another way to thumb their nose at the government. In September 1978 they robbed Rs. 30,000 from the manager of the Kopay Multi Purpose Co-operative Society. In December came a bigger heist. The People's Bank branch at Nallur was raided and Rs. 1,120,000 taken by six youths who came on bicycles. The two policemen on guard duty at the bank were killed in the shoot-out. The spiral of violence continued into the next year. On 25th March 1981, nearly eight million rupees belonging to the Neerveli People's Bank Branch was robbed by a gang of armed militants. Two police constables escorting the money were killed on the spot while the gang escaped unscathed.[7]

By now there were several militant groups leading the charge against the Sri Lankan state. Apart from the Liberation Tigers of Tamil Ealam (LTTE), there was also the Tamil Ealam Liberation Organisation (TELO), People's Liberation Organisation of Thamilealam (PLOT), Ealam People's Revolutionary Liberation Front (EPRLF) and Ealam Revolutionary Organisation of Students (EROS). Velupillai Prabhakaran had renamed the Tamil New Tigers the LTTE in 1976 while TELO had originated from a group of militants around two leaders, Nadarajah Thangathurai

---

[7] Sarath Munasinghe, *A Soldier's Version*, Colombo: author publication, 2000, p. 54.

and Selvarajah Yogachandran or "Kuttimani" in 1979. The PLOT under Uma Maheswaran was the result of a split within the LTTE. Uma quarrelled with Prabhakaran in 1978 and formed a separate group. EROS had also emerged in the 1970s while the EPRLF was an offshoot of the EROS.

The strategy of these diverse groups at this stage, was little more than intimidation of the government and also of collecting weapons and funds to build up their organisations. They were picking off individuals and robbing establishments that were easy targets for small groups of armed men. Whenever possible, arms were being taken from their victims as was the case with Bastiampillai's escort. Their actions were very much those of the terrorist and bandit, albeit in the service of a political cause.

But soon the violence escalated, following a familiar trajectory in most armed insurgencies. On 27th July 1981 the PLOT announced the budding guerrilla aspirations of the rebels by raiding the Annacottai police station. The militants drove in a hijacked van, shot dead a constable and fled with a sizeable haul of arms and ammunition.[8] In October 1982 the LTTE decided to ratchet up the violence in a different – and bolder fashion. The target chosen was the Chavakachcheri police station. A group of eight hardcore LTTE cadres led by one of their seasoned veterans Seelan arrived at the police station in a hijacked mini bus, opening fire as they approached. In the ensuing fire-fight which is said to have lasted nearly 15 minutes, two more policemen including a police driver were killed and at least two others wounded. The attackers escaped with a haul of weapons and ammunition.[9]

---

[8] Ibid., p. 61.

[9] M. R. Narayan Swamy, *Inside an Elusive Mind: Prabhakaran*, Delhi: Konark Publishers, 2003, p.73, Rajan Hoole (et al), *Broken Palmyra:*

In October 1981 the militants also took on the army for the first time, killing two men of the Engineering Services Regiment. The men were part of a four-man detail purchasing building materials at a shop on Kankesanthurai Road in Jaffna when four armed militants road up in bicycles, shot dead two of the men waiting by their vehicle and escaped with the rifle of one of them. The army lost another soldier in 1983 when on 18 May, 1983 during a by-election, the LTTE launched an attack on the polling booth at Sivaprakas Maha Vidyalayam at Kandemadam. The booth was guarded by five soldiers of the recently raised Rajarata Rifles regiment and five armed policemen. In the attack launched with firearms and grenades, Corporal Jayawardena died and one private and two policemen suffered injuries. The Tigers escaped with the dead soldier's T56 rifle and two magazines. Elsewhere, the Air Force also came for some punishment on the ground. Just days after the attack on the polling booth cadres of the PLOTE attacked an Air Force detail on a provision buying mission at the Vavuniya market. The militants shot at the airmen and then during the ensuing gun battle threw a grenade into the rear of the jeep killing two airmen instantly.[10]

In September 1982 in an ominous sign for the security forces, the Liberation Tigers tried their hand a new weapon: the landmine. The landmine was a potentially deadly new addition to the militant's arsenal. It could do what a rifle or grenade could not do: blow up a vehicle full of soldiers. The first attack however, was far from successful. The mines used were crude home-made devices, nothing more than caste iron cylinders packed with pieces of metal and explosives. The blast was to be triggered off by a Honda generator connected to the mines by

---

The Tamil Crisis in Sri Lanka, an Inside Account, Ratmalana, Sri Lanka: Sri Lanka Studies Institute, 1992, p. 38.

[10] L. M. H. Mendis, *Assignment Peace: In the name of the Motherland*, Nugegoda, Sri Lanka: Author Publication, 2009, pp. 29-30.

means of a wire. The attack was launched on 29th September 1982 against two navy jeeps escorting three water bowsers on the Ponnalai causeway that connected the Karaithivu island to the mainland. The causeway was mined at four places near its southern end and the militants took position in the scrub that grew on the sand mounds in the lagoon. As the army vehicles approached a militant activated the mines but only one mine exploded and that too prematurely. The militants escaped to a waiting mini bus leaving the naval escort too stunned to retaliate.[11]

Undaunted, they made another attempt, this time on 4th March 1983 just 4 km south of Elephant Pass. The attack was made on an army truck carrying breakfast for a group of soldiers deployed at the Kilinochchi police station. The truck was escorted by four soldiers. The Tigers buried two mines in the road near Umayalpuram Kovil and they were to be detonated using a lorry battery. Again, the blast went off prematurely, with the truck was about 50 meters away.[12] The soldiers fled back to the Elephant Pass camp and seeing the approach of another convoy from Kilinochchi to the south, the militants too took to their heels.

The reality was that despite their deadly attacks the militants were still learning their trade. This was clearly shown by the abortive attempts with landmines. They were experimental attacks carried out by amateurs. The militants were also working with limited resources. During the early days their weapons were dominated by a few revolvers and shot guns. They had captured two sub machine guns, the first when they killed Inspector Bastiampillai in April 1976 [13] and the second when they robbed the Tinneveli People's

---

[11] 'The Pirapaharan Phenomenon', Chapter 30,
http://www.sangam.org/articles/view/?id=212
[12] Munasighe, *A soldier's Version*, p. 85.
[13] Narayan Swamy, *Inside an Elusive Mind*, p. 47.

bank branch.[14]   But it was not an armoury that caused too many concerns to the army; Captain Sarath Munasinghe who arrived in Jaffna in January 1979 to carry out intelligence gathering operations felt a World War II vintage .38 revolver was "more than sufficient" for his security at the time[15]. The only significant purchases the Tigers had made were a G3 rifle for Prabhakaran from a retired Indian army officer for 3000 rupees and a .38 revolver for 300 rupees in 1980.[16] To be sure, the militants' arsenal was growing with their attacks as they were able to prise weapons from their victims. The eight men who stormed the Chavakachcheri police station armed with one SMG, one G-3, one repeater rifle, two revolvers and grenades had left the station carrying an additional two sub machine guns, one .38 revolver, nine .303 rifles and nineteen repeater shotguns, the tigers' biggest arms haul to date.[17] The attack on the polling booth in May 1983 also gave them their first T56 rifle. Still, this was not a potent arsenal. According to military intelligence in July 1983 the LTTE possessed a paltry collection of weapons: six submachine guns, five .38 revolvers, ten SLRs, two .45 revolvers, eleven .303 rifles, one .30 carbine, one T56 rifle and twenty-four repeater shot guns[18]. The other groups were even less well armed, a few SLRs and sub machine guns taken from the police and army being the most lethal weapons in their armouries. The experiments with landmines had given the Army a fright but they had been failures.

---

[14] 'The Pirapaharan Phenomenon', Chapter 16, http://www.sangam.org/articles/view/?id=45
[15] Munasinghe, *A Soldier's Version*, p. 48.
[16] Narayan Swamy, *Inside an Elusive Mind*, p. 58.
[17] 'The Pirapaharan Phenomenon', Chapter 30, http://www.sangam.org/articles/view/?id=212
[18] Munasinghe, *A Soldier's Version*, p. 10.

Then, in July 1983 the LTTE made the boldest move yet, ambushing a military convoy. The ambush was carefully planned and well executed. The place chosen for the attack was Tinneveli, a village about 2 km outside Jaffna town. It was on the way taken by the routine night patrol that plied between Madagal and Gurunagar army camps. The militants expected to hit the patrol on its return journey from Gurunagar to Madagal. They had buried two landmines about two meters apart. The plunger which had been robbed from the Kankesanthurai Cement plant just over two weeks ago was placed on the roof of an adjoining shop. The roof was flat with a half-wall around it. While two cadres took position on the roof with the detonator, the rest concealed themselves behind walls on either side of the road.

The patrol consisting of a jeep and a truck ran into the ambush around 11.20 pm. The first landmine exploded under the jeep killing all four occupants instantly. The second mine opened a huge crater in front of the truck which came to a halt only to be assailed by gunfire from all sides. Within minutes it was all over. By the time army units rushed to the scene the Tigers had vanished with the weapons of the soldiers. Twelve soldiers lay dead. Another soldier who was wounded died on the way to the hospital. It was the biggest single loss suffered by Sri Lankan security forces in the war against Tamil militants up to date.[19]

The Tinneveli attack was militarily significant for a number of reasons. Not only did the militants finally get a landmine to go off at the right moment but they also deployed an unprecedented volume of firepower in the ambush. Military officials estimated that around 26 Tigers had taken part in

---

[19] T. D. S .A. Dissanayake, *War or Peace,* pp.34 - 6, Munasinghe, *A Soldiers Version*, pp. 4 - 6.

the attack armed with an assortment of weapons including SLRs, repeater shot guns, hand grenades and the solitary T56 captured from the polling booth a few weeks ago, undoubtedly the bulk of the arsenal possessed by the Tigers at this time. It was a classic guerrilla attack: bringing overwhelming firepower to bear on the target in a quick surprise assault. And the target was not a lightly manned outpost or police station but a heavily armed military patrol. It was a significant departure from all the previous operations. The haul of weapons was also rich; a dozen SLRs, a shot gun and a small pile of ammunition.

2.3 Mother India

The news of the ambush at Tinneveli led to widespread anti-Tamil rioting in Colombo and several other cities. Hundreds of Tamils were killed, thousands of Tamil-owned residences and businesses burned and looted and thousands of Tamils made homeless. In many places the riots were well organised and often the police and the military stood by idly without stopping the violence and in some cases even took part in it. One of the more gruesome and disturbing episodes during the carnage occurred within the Welikade jail in Colombo where a number of Tamil prisoners including several militants were brutally massacred by Sinhalese inmates.[20]

July riots were a watershed in the Tamil insurgency. The rapid expansion and growing sophistication of the militants was one key consequence of the mayhem in July 1983 that made it a turning point in the war. The mobs in Colombo and other cities, while "punishing" Tamils by burning and looting Tamil

---

[20] L. Piyadasa, *Sri Lanka: the Holocaust and After*, London: Marram Books, 1984, pp. 80-84, S. Sivanayagam, *Sri Lanka: Witness to History, a Journalist's Memoirs 1930 - 2004*, London: Sivayogam, 2005, pp. 245-72. Among the militants killed in the prison massacre were Thangathurai and Kuttimani.

properties and killing defenceless Tamils had also fuelled the anger and insecurity of a people. Hitherto many Tamils outside the North and the East had remained somewhat aloof from the militancy and the militants were being treated with tolerance and even with some apprehension. Even in Jaffna which was politically conservative, the militants' need to resort to violence was understood but not necessarily approved. But after the riots few Tamils in Sri Lanka felt secure under a Sinhalese-dominated government and almost overnight, the militants turned into saviours, defenders of the dignity and lives of the Tamils. Tamil youth who were smarting from the indignities of July '83 now naturally turned to the militants as an avenue of revenge and hope. Within days of the riots thousands of eager Tamil men and boys were seeking enlistment with the Militant groups.

But having thousands of eager recruits does not make an army. They need training, weapons and money. These came from across the seas, from neighbouring India.

A mere 22 miles of shallow sea separates Tamil Nadu from Jaffna. Across this narrow Palk Strait the people of Tamil Nadu and Jaffna have shared a culture and language for centuries. Visitors from India instantly recognised the similarity, a Bombay journalist quipping in 1984 that "one may be pardoned for mistaking Jaffna for a medium sized town in Tamil Nadu", before going on to observe rather pointedly that although Jaffna's body was in Sri Lanka its soul was in Tamil Nadu.[21] There was much truth in this. Jaffna Tamils looked up to south India for cultural nourishment, idolising south Indian movie stars and following sub continental fashions and trends closely. Even the informal economy of Jaffna was closely tied to Tamil Nadu, towns like Velvetithurai bustling centres of smuggling between Sri Lanka and India. Many Tamils regularly travelled

---

[21] Quoted in Sivanayagam, *Sri Lanka: Witness to History*, p. 236.

back and forth between Jaffna and the cities in Tamil Nadu for business and pleasure.[22] For many Tamils in the north, India was a closer home than Sri Lanka.

It was only natural then that the Tamil struggle in Jaffna aroused the sympathy of many Tamils in Tamil Nadu. Local politicians like P. Nedumaran followed Sri Lankan Tamil politics closely and showed a keen interest in the direction of the militants' struggle. Understandably, Tamil Nadu developed as a safe haven or a 'rear base' for Tamil militants from an early time. Guerrilla leaders like Prabhakaran frequently crossed over to Tamil Nadu to evade arrest. When the army came down heavily on the militants it was to Tamil Nadu that many militant leaders fled. There, in camps set up in Tiruchirappalli and Madurai, they underwent training, under the guidance of some retired Indian officers who spoke Tamil.[23] And when LTTE operatives were wounded they were shifted to Madras for medical attention.[24]

Still, until July 1983 South India was merely a convenient 'rear base'. But after the riots in 1983, India came to play a bigger role in the Tamil struggle. Now it became a source of training, weapons, money and patronage at the highest levels of the Indian administration. The Tamil insurgency now became part of India's strategic considerations.

Naturally the worsening situation in Jaffna aroused much concern in Tamil Nadu. The riots in July '83 and the flood of refugees also raised tempers. However, not only the Tamils in India had an interest in the plight of the Tamils in Sri Lanka. The island's geographical proximity to India and the cultural ties between Tamil Nadu and the northern province of Sri

---

[22] Cyril Ranatunge, *Adventurous Journey: from Peace to War, Insurgency to Terrorism*, Colombo: Vijitha Yapa, 2009, p. 75.
[23] Narayan Swamy, *Inside and Elusive Mind*, p. 57.
[24] Ibid., p. 72.

Lanka meant that events in Sri Lanka and Sri Lanka's strategic decisions would be closely monitored by India, led by her strong willed Prime Minister Indira Gandhi. After the victory of the United National Party in 1977 under J. R. Jayawardena who had a reputation as a friend of the West, this scrutiny became closer. India, with her close relationship with the Soviet Union, was predictably irked by the Jayawardena regime's pro-western stance highlighted by Sri Lanka's support for Britain in the Falklands War. Western countries were also pouring money into Sri Lanka's development projects extending their influence over the island. These concerns grew when India began to suspect that Sri Lanka was favouring an American consortium for the contract for repairing and restoring the WW II vintage oil tanks in Trincomalee. America was a strong supporter of India's traditional rival Pakistan and an American foothold in Trincomalee which was one of the finest natural harbours in the world was too much for India to stomach. Last but not least, the central government could not ignore the public opinion in Tamil Nadu which was firmly behind the Tamils in Sri Lanka.[25]

Increasingly resentful of Sri Lanka's cosying up to the West, India was keen to show its tiny neighbour that it was not free to ignore India's interests in the region. The riots in July '83 made it imperative. The disturbances drove tens of thousands of Sri Lankan Tamils to Tamil Nadu, bringing with them horror stories that incensed public opinion in South India. The situation in Sri Lanka was threatening to become a problem for India itself, something which India could not overlook anymore. India had to put her foot down, fast, and the boot was provided by the militants.

---

[25] Thomas A. Marks, 'Insurgency and Counterinsurgency', *Issues and Studies*, August 1986, pp. 92-5.

To India, the militants offered an excellent way of twisting Sri Lanka's arm, to remind that it ignored India's interests in the region at great cost to itself. According to author and analyst Narayan Swamy, as early as 1981 Indira Gandhi instructed the Research and Analysis Wing (RAW), India's intelligence organ, to reach out to Prabhakaran to find ways of using the militants for strategic leverage against Sri Lanka, to obtain information about western involvement in the island.[26] The efforts intensified after July 1983. RAW officials began making contact with the militant groups again, sending out feelers to gauge their willingness to embrace India's generosity.

Naturally, the militants were thrilled. They were somewhat suspicious at first but they soon realised that the offer was genuine. And it could not have been better timed. The riots had outraged The Tamils in Sri Lanka driving them to the militants' fold. Earlier the militants had been treated with sympathy and even admiration by the people in the north but the sympathy had not often extended to joining their ranks. Many Tamils had still believed that politicians rather than militants should lead the struggle. But after 1983 the view changed almost overnight with many a young Tamil now feeling that the armed struggle was the only way to achieve freedom and dignity. As refugees arrived in Jaffna with tales of horror the militants' ranks swelled with angry young men eager young men impatient to get back at the state they held responsible for humiliating and murdering their people.

The militants accepted the Indian offer. Soon boatloads of young Tamils were leaving for training in India, braving the choppy seas in little rickety boats to learn what the Indians had to offer.

---

[26] Narayan Swamy, *Inside an Elusive Mind*, pp. 67-8.

And the Indians had much to offer. By September-October 1983 training had begun in earnest in south and north India. In South India training camps had sprouted in the districts of Tiruchirappalli, Madurai, Thanjavur and in Madras while in the north they were established at Dehra Dun and near New Delhi and also in Uttar Pradesh. Some of the camps in the south were run by the guerrilla groups themselves while the authorities turned a blind eye to them. For example PLOTE maintained a number of camps in Thanjavur while the LTTE established a camp in Madurai. Others were run by RAW itself. RAW provided the instructors, many of whom were retired Indian Army officers.[27]

The recruits were trained in a range of skills important to the guerrilla, including the use of assault rifles, mortars and rocket launchers and in reading maps and handling explosives.[28] In some camps even the use of revolvers and 12 and 16 bore shot guns was taught.[29] They were also put through gruelling physical drills to condition them to the rigours of combat. The Indians also provided the armed groups with an assortment of weapons which included SLRs, RPGs, AK-47s, sub machine guns and light machine guns.[30]

By 1987 India is said to have trained thousands of militants. TELO was the group to benefit the most, having sent the largest number of recruits. Upon completion of their course the trained militants were brought down to Madras by bus from where they were put on boats and returned to Jaffna. PLOTE is even said to have trained a few Sinhalese supportive of their cause.[31]

---

[27] Rohan Gunaratna, *Indian Intervention in Sri Lanka: the Role of India's Intelligence Agencies*, Colombo: South Asian Network on Conflict Research, 1993, pp.39-43, Narayan Swamy, *Inside an Elusive Mind*, pp. 96-7.
[28] Gunaratna, *Indian Intervention*, pp. 39-43.
[29] 'Inside a TELO Training Camp', *Weekend*, 11. 11. 84, p. 1 and p. 5.
[30] Gunaratna, *Indian Intervention*, p. 140 and p. 153
[31] 'Top Secret Camp for Terrorists', *The Island*, 5. 10. 86, p. 9 and p. 15.

Indian training boosted the morale and battle readiness of the militants. However not everybody was satisfied. According to Anton Balasingham, the LTTE's theoretician, LTTE leader Prabhakaran was disappointed with the level of Indian support. The Indians clearly did not want the militants to acquire a sophistication that would enable them to defeat the Sri Lankan forces. They were trained in the use of small arms that would have helped them to fight the Sri Lankan military to a standstill, not in modern sophisticated weapons systems that would have enabled them to defeat their enemy decisively.[32] It was a grudge that Prabhakaran would bear for a long time.

Frustrated by the Indians' chicanery, the LTTE began exploring alternative avenues for augmenting their arsenal. The problem however was that while the Indians were offering their meagre supplies free of charge, independent procurement of weapons required a substantial treasury. This problem was solved by the generosity of M. G. Ramachandran or MGR, the actor- politician in Tamil Nadu. MGR gave the LTTE two million Indian Rupees, a precious financial windfall that raised the rebel group's stocks considerably. "The close and intimate relationship between the Tiger leader and MGR and the firm political support and huge financial assistance provided by this legendary figure became the cornerstone for the development of the LTTE," observed Balasingham later, recognising the pivotal role of this contribution.[33] The Tigers now had the money to purchase the arms they felt they were denied by India.

---

[32] Anton Balasingham, *War and Peace: Armed Struggle and Peace Efforts of Liberation Tigers*, Mitcham: Fairmax Publishing Limited, 2004, p. 61.
[33] Ibid., p. 62

With plenty of spending money, Prabhakaran now set his sights on developing his own arms procuring establishment. This was set up under K. Pathmanathan - or K.P. as he came to be known - who travelled the world shopping for the LTTE in arms bazaars of Asia, Europe and the Middle East. A shipping line was established for the purpose. The first consignment of arms arrived in south India without the knowledge of the Indians in 1984. The weapons were promptly dispatched to Sri Lanka by boat[34].

India continued to train the militants. But it was now becoming increasingly difficult to keep it a secret. In March 1984 the Sri Lankan Army raided a suspected militant hideout in Point Pedro in Vadamarachchi and took some suspects in for questioning. Some of the youth claimed that they were trained in India.[35] The revelation worried Sri Lankan government. It was one thing to face a band of home-grown militants but to fight a guerrilla army trained and armed by India with secure bases on the subcontinent was a daunting prospect. Soon word began to seep out about the many camps that trained militants on Indian soil. In March 1986 the *South* magazine even went so far as to publish detailed list of camps in India.[36] As the existence of the camps became public knowledge the Sri Lankan government requested India to close them down but India repeatedly denied their existence. There were no terrorist camps in India the Indian government said firmly. There were only refugee camps. Indian Prime Minister Rajiv Gandhi even invited journalists to visit India and check for themselves.[37]

---

[34] Narayan Swamy, *Inside an Elusive Mind*, pp. 108-110.
[35] Edgar O'Ballance, Cyanide War: the Tamil Insurrection in Sri Lanka 1973-88, London: Brassey's, 1989, p. 35.
[36] 'Military Training in Tamil Nadu and India', *The Island*, 5. 10. 86, p. 9.
[37] 'Is Rajiv Gandhi Being Misled?', *The Island*, 5. 10. 86, p. 9.

As the militants gradually grew in numbers and confidence they opened camps on Sri Lankan soil itself. The various militant groups now vied with each other to recruit and train fighters. They enjoyed showing off their growing strength especially to visiting journalists. In July 1985 a reporter from Colombo was taken to watch about 75 LTTE guerrilla trainees going through mock battles in their camp which was on the outskirts of Jaffna. An EPRLF leader told the same reporter that their recruits were trained for 10-14 weeks before they were considered fit to "face the enemy".[38] Another reporter from the south of Sri Lanka watched around 175 LTTE recruits going through rigorous drills that included scaling walls and creeping under barbed wire with weapons in hand and taking part in "mock commando style assaults". The training grounds were a public park only 5 miles from the Jaffna city centre.[39] While these camps were in the peninsula itself, the jungles deep in the Vanni also provided ideal grounds for building the rebel army. A journalist of the Associated Press was treated to the spectacle of about 150 recruits going through a 'gruelling 3 and a half hour workout' in one such camp. Ominously, most were teenagers, some as young as twelve. The commander of the camp explained that the recruits were trained for six months. He claimed there were three Tiger camps in the Vanni, training about 400 cadres, the one visited by the journalist being the largest.[40]

Along with their training the militants also began to intensify the campaign against suspected informants in an effort to bring the population into line. For the militancy to succeed quislings had to be stopped and stopped brutally.

---

[38] Faizal Samath, 'Terrorist Training Camp at Jaffna', *The Island*, 14. 7. 85, p. 1 & p. 2.
[39] Dexter Cruez, 'Within the Jaws of the Tiger', *Weekend*, 26. 10 . 86, p.8.
[40] 'Inside a Tiger Training Camp', *The Island*, 8. 7. 85.

The first half of 1984 alone saw at least 40 assassinations of suspected informants by militants. The victims' bodies were usually tied to lamp posts with placards proclaiming their treacherous activities.[41]

While the informants perished the militants' ranks were expanding rapidly. The years 1984-85 were the peak in the militants' popularity, boosted by the promise of Indian support and the memories of July '83. According to the analyst the late Dharmaratnam Sivaram, the total number of militants who had basic military training before July 1983 was 800. In 1984-85 this has grown to 44,800. According to him PLOTE had the largest number at that time with 6,000 cadres under training in its camps in South India and around 12,000 cadres in the camps in the North-East of Sri Lanka. TELO had 4,000 cadres under training in its camps in South India and 2,000 in the North-East. EPRLF had about 7,000, including 1,500 girls, in South India and the North-East. The LTTE had less than 3,000 cadres and EROS 1,800 cadres. The balance belonged to the smaller militant groups.[42] This appears to be very optimistic but not fantastic when one considers that Sivaram's numbers only relate to those who received training, not necessarily those who were armed.

2.4 The Tamil 'Offensive' 1984-85

If the mobs in the south had thought they had taught the Tamils a lesson in July 1983 they can be pardoned. After the riots in 1983, violence in the north abated for a while. Despite the immense boost given by the riots, Tamil militants took time to return to the offensive after July

---

[41] O'Ballance, *Cyanide War*, p. 39.
[42] 'Taraki' (D. Sivaram), 'The cat a bell and a few strategists', *Sunday Times*, 20. 4. 97, p. 7.

1983. There were a few acts of vandalism and terrorism, such as the burning of buses but no major incident occurred until late 1984. But this was not due to any sense of chastisement felt by the Tamils. The focus had now shifted to India where the militants were being trained.

But still there were intermittent rumblings. On 1st January 1984, a police vehicle was attacked in Point Pedro, killing two policemen. This was the first attack on the police or army since July 1983.[43] In April 1984 an army truck driving past a church was attacked with a bomb rigged to a car wounding 14 soldiers. Around the same time bombs were hurled at the Point Pedro police station causing extensive damage.[44] If the North was calm it was an uneasy calm indeed.

Then in August, violence broke out in a sustained spell changing the tone of the war. The trigger was provided by a confrontation between the navy and the militants at sea. On the night of the 4th August Sri Lankan naval craft came into contact with a rebel boat off the coast of Velvetithurai. In the ensuing exchange of fire at least two sailors were killed, the first naval casualties in the war. Stung by the loss, the security forces immediately launched a search operation on the coast. It only led to more casualties. During the search the guerrillas set off a landmine under a police jeep killing an Assistant Superintendant of Police and wounding several soldiers. The following morning a military convoy including armoured cars was employed in taking the wounded men to the Palali airbase to be flown to Colombo. The convoy came under repeated attack by the rebels who had set up road blocks. It was only

---

[43] O'Ballance, *Cyanide War*, p. 35.
[44] 'Violence in Jaffna: Five killed', *New Straits Times*, 11. 4. 84, 'Nine Die in Sri Lanka Riot', *The Australian*, 12. 4. 84.

after an armoured car fired its 76 mm cannon that the convoy was able to move ahead.[45]

The same night the militants attacked the Oddusuddan police station in the Mullaitivu District, killing an inspector and a policeman and capturing a large haul of weapons and ammunition. Some reports even claimed the attackers were dressed in military fatigues. The police station was left badly gutted. The following morning the Police Superintendant of Vavuniya walked in to his office and sat down at his desk only to be blown up by a booby-trapped bomb. On 10th August two branches of the People's Bank in Jaffna were raided and safes stolen.[46] The following day the violence moved south-westward. A landmine in Mannar on the Mannar - Pooneryn road killed six soldiers and wounded another. It was the biggest loss suffered by the army since July 1983.[47] Three days later the Kayts and Velvetithurai police stations came under attack.[48]

Alarmed by the sudden spike in violence the government claimed that the guerrillas had launched a major offensive. This may have been somewhat exaggerated. Brigadier Nalin Seneviratne, the man in charge of military operations in the north was less sanguine than the politicians about the 'offensive' but he did not dismiss it as the work of a handful of youths either. The army, he noted, was now dealing with a different kind of enemy than before. "The method of their attacks on security forces seems to be more sophisticated in the way of equipment as well as training. The numbers involved have also increased," he pointed out.

---

[45] 'Sri Lankan Separatists Launch Attacks', *New Straits Times*, 7. 8. 84.

[46] 'Sri Lankan Guerrillas Raid Two More Banks', *Singapore Straits Times*, 10. 8. 84.

[47] 'Six Servicemen Die in Ambush', *Daily Mirror*, 13. 8. 84.

[48] Hana Ibrahim, 'Fresh Attacks on Police Stations in North Repulsed', *Daily Mirror*, 15. 8. 84.

"Where as previously 8 or 10 were coming in bicycles now they hijack vehicles on the road and come with two or three vans or lorries. It could be 20 or 30 terrorists at a time. And to carry out an operation and swing their people around they are using radio transmitters. Their command and tactics seem to be well controlled. They know what they are doing. Before it was hit-and-run. Now they're standing and giving the security forces a fight." The Brigadier also noted the growing sophistication in the guerrilla arsenal which now included sterling submachine guns, self loading rifles as well as repeaters and shot guns and more professionally made grenades and hand bombs.[49]

The writing on the wall was clear. The militants trained and armed by India were back. And they were back with a vengeance.

Violent confrontations between the security forces and the guerrillas now began to occur with alarming frequency. The security forces were coming in for a steady hammering from the rejuvenated and buoyant guerrillas who took every opportunity to kill, maim or even simply annoy them. Gone were the days of amateurish experimenting with landmines. The guerrillas were steadily becoming deadly perfectionists in the lethal art of blowing up vehicles. The roadsides in Jaffna were littered with the wreckages of military vehicles and the morgues and hospitals full of military and police casualties to prove it.

On 2nd September 1984, six police commandos died in a landmine attack at Tikkam near Point Pedro.[50] A week later the guerrillas struck outside the peninsula, setting off a mine in

---

[49] Eric Silver, 'Tamils Main losers in Unwinnable War', *Canberra Times*, 18. 8. 84.
[50] 'Six Commandos Killed', *Saturday Review*, 8. 9. 84, p. 1.

Mullaitivu on the east coast, killing six soldiers.[51] Another mine blast took the lives of nine soldiers on the Achchuvely - Vasavilan road in Jaffna on 1st November and the following day another six soldiers died on the Thondamanaru- Palali road not far from the main army camp at Palali[52]. On 5th December a mine blast in Murunkan in Jaffna killed one army driver and wounded six soldiers.[53] The list went on and on.

On 19th November 1984 the militants raised the stakes with a deadly blow. That day Colonel A. Ariyapperuma, the commander of the Sri Lanka Army's northern command was killed by a landmine explosion at Tellipalai in Jaffna. The Colonel was on his way to investigate the blasting of a culvert a few days prior to that. The armoured personnel carrier he was travelling in was thrown off the road. The mine had been detonated from behind two trees in the neighbouring scrubland.

Then on 23rd November 1984 the militants showcased their growing boldness in one of the most daring and successful attacks of the war to date. This was the devastating attack on the Chavakachcheri police station.

The attack came at an unusual time, in the afternoon, around 2.30 pm. Using a young boy as a decoy, scores of young men and boys, some of the former in military-style uniforms and some of the latter in school uniform, attacked the police station with firearms and grenades. A few sticks of gelignite were thrown in the barrack room and the guard room and detonated, bringing the building down in a heap of masonry, trapping the policemen inside. As reinforcements rushed to the scene they were ambushed injuring an officer and six soldiers in the relief party. When

---

[51] 'Six Soldiers Killed in Rebel Ambush', *Malaysian Straits Times*, 11. 9. 84.
[52] 'Pirapaharan Phenomenon', Chapter 25,
http://www.sangam.org/articles/view2/?uid=645
[53] 'Tamil Rebels Strike Convoy', *The Australian*, 6. 12. 84.

the army finally arrived it was all over. All that was left to do was to pick up the pieces. When the debris was finally cleared, which took till the following afternoon, they recovered 29 bodies, 24 of them belonging to policemen.[54]

The attack on Chavakachcheri demonstrated the growing sophistication and prowess of the militants. What were impressive were the numbers involved and the coordination of the various elements of the assault. The participation of school children was noteworthy, revealing the growing hostility to the security forces in the community. This was a people in revolt, the young leading the charge.

On 19th January 1985 came another shocking attack. The militants blew up the Yal Devi train from Jaffna to Colombo as it was pulling out of Murunkandi south of Kilinochchi, under heavy military escort. The militants hiding in the surrounding jungles opened fire on the soldiers who were struggling in the wreckage. When the shooting stopped, 39 lay dead, 28 of them soldiers.[55]

Then the militants attacked an army camp for the first time. On 9th February, 1985 a large group of armed militants attacked the army base at Kokilai on the eastern coast. The small garrison repulsed the assault but it showed the growing sophistication of the enemy. The rebel cadres had used RPGs for the first time in an attack and the bodies of the 14 militants found dead outside the camp perimeter were clad in military style uniforms. They also carried food packets, water bottles

---

[54] Sri Lanka (Ceylon) News-Letter published by the High Commission of the Democratic Socialist Republic of Sri Lanka in Canberra, 10th December 1984, p. 10

[55] 'Yal-Devi Blast: Death Toll 39', *Sun*, 22. 1. 85, Aruna Kulatunga and Premalal Wijeratne, 'The Train Tragedy', *Sun*, 22. 1. 85.

and medicine. Some even had night vision glasses.[56] The camp had not been attacked by a rag tag band of desperadoes. The enemy was well trained, armed and organised.

By now the militants were clearly aiming at shifting to a sustained guerrilla campaign. "We strongly believe in the guerrilla mode of warfare" declared Anton Balasingham in Madras, revealing the evolving strategy of the rebels. "The essence of the tactic is to get your arms from the enemy, launch unexpected attacks and thoroughly demoralise the enemy before the step of declaring full scale war."[57] This, the militants seem to be doing very well. Another militant identified as A. S. Skantha was quoted as saying that the militants were now looking at moving beyond the hit-and-run stage, focusing on defeating the army militarily. He boasted that the militants were in control of 80-85 percent of the countryside in the north while the army controlled only the towns.[58] Such claims were probably overly optimistic at this stage but they reflected the ambitions of the increasingly confident guerrillas.

And the violence continued, assuming more alarming proportions. In April 1985 they made an even more audacious bid. On 10th April a large group of LTTE guerrillas surrounded the Jaffna police station less than a mile from the Jaffna fort military encampment. They cut off the electricity to the city first and as the town plunged in to darkness the guerrillas launched a devastating attack on the police station. The militants claimed to have captured "200 weapons, more than 500 sophisticated hand grenades, 20,000 bullets, gas guns and pistol guns" without suffering a single casualty. Several policemen who had surrendered were to be released "after they

---

[56] Malinga H. Guneratne, *For a Sovereign State*, Ratmalana, Sri Lanka: Sarvodaya Book Publishing Services, 1988, pp. 239-40.
[57] S. H.Venkatramani, 'Battle Lines', *India Today*, 15.12.84.
[58] 'Terrorists planning Easter Invasion', *Daily News*, 28.2.85.

have gained a clear knowledge of the struggle." In the South the government's organ *Ceylon Daily News* painted a totally different picture of the attack claiming that the "terrorist force loss at least 21 fighters." The report however, betrayed the reality when it revealed that the double storey DIG's building was flattened and that the bodies of three policemen were being dug out of the rubble.[59] In the coming weeks the militants pressed their offensive attacking Army camps and police stations all over the north. The Karainagar and Gurunagar camps on the Peninsula and Kokavil and Kilinochchi camps in the mainland were assailed.[60] For the attack on Kokavil in May the militants also unveiled a new weapon. The attackers, in this instance the TELO, approached the camp in two hijacked bulldozers protected by metal plating.[61] The improvised 'tanks' were stopped in their tracks but it marked an ominous escalation in the rebel offensive. On 10th May the Mannar police station too suffered a particularly serious assault. The attack flattened the police station and killed five policemen while four were reported 'captured'[62].

Clearly, the guerrillas were beginning to flex their growing muscle. They were making a habit of hitting heavily defended enemy positions. Even though most such attacks were not as successful as the attack on Chavakachcheri or the Jaffna police station they underscored the burgeoning confidence of the militants. And the failure to demolish or over run enemy camps or compounds did not deter them from making their presence felt in other ways. By late 1984 rebel bunkers were coming up within close proximity to the military camps, penning the soldiers in. Landmines were

---

[59] *Ceylon Daily News*, 13. 10. 85, p. 14.
[60] Mendis, *Assignment Peace*, p. 49, *Sri Lanka Army 50 years on*, Colombo: Sri Lanka Army, 1999 , p. 428. *Sri Lanka Army* erroneously dates this attack to 1984.
[61] *Sri Lanka Army,* p. 428.
[62] 'A New Spiral of Violence', *Asiaweek*, 24 .5. 85.

buried around the camps and militants mounted guard round the clock watching the movements of the enemy. The army was steadily becoming a prisoner within its camps.[63]

These were alarming developments for the government; but more shocks were on the way. Early in the morning on 14th May, a group of about 20 LTTE guerrillas hijacked a bus from the Puttalam depot and drove to Anuradhapura, a city holy to Buddhists. When they arrived at the Central bus stand, the place was already filling with people bound for work. An eyewitness described how three men dressed in military uniforms got out of the bus and surveyed the area. Suddenly a man came riding a motorcycle giving the thumbs up signal. The men in the bus opened fire at the people at the bus stand. The bus then drove slowly through the streets, the guerrillas firing away at the fleeing people. The bus made its way to the sacred Sri Maha Bodhi[64] where they continued their killing before hijacking another bus and speeding back towards Puttalam. On the way they fired at the Nochchiyagama police station and when they arrived at the Wilpattu National Park, shot dead 24 of the employees and forced one to be their guide. At the end of the rampage, 148 people lay dead including 120 killed at the Sri Maha Bodhi. The latter included three monks and 2 nuns. Over a hundred were injured. The ghosts of 1983 had come to haunt the Sinhalese.[65]

The war had changed dramatically since the days before the July riots. The brutal reaction to the ambush at Tinneveli

---

[63] 'The Pirapaharan Phenomenon', chapter 25,
http://www.sangam.org/articles/view2/?uid=645
[64] The Sri Maha Bodhi is an ancient fig tree (Ficus Religiosa) in the ancient city of Anuradhapura. It is said to have grown from a sapling of the fig tree under which the Buddha attained enlightenment. It is sacred to the Buddhists in Sri Lanka.
[65] 'Tamil Terrorists Kill 150, Wound 300 in Sacred City Attack', *The Australian* 15.5.85, 'Tamil Killings a Reprisal for Earlier Village Deaths', *The Australian*, 15. 5. 85.

had produced an even bigger reaction. The government now had more "terrorists" to deal with than ever before and they were proving to be deadlier too. Instead of the old hit and run attacks launched by individuals or small groups with revolvers, grenades and shot guns, now the militants were carrying out operations involving dozens of men, armed with automatic weapons, mortars and landmines. The targets were no more the lone policeman or the slackly guarded outpost. The militants were boldly taking on well guarded and manned installations. They were launching more and more audacious attacks that challenged the security forces' hold on the north, attacking them everywhere, whether they were on patrol in the open or whether they were in their seemingly more secure bases. What was even more chilling was the willingness of the militants to take the war to Sinhalese dominated areas. It jolted the government and the Sinhalese people, awakening them to the deadliness of the struggle that was evolving.

For the militants these were heady days. They had risen as a people and brought the state's armed forces to their knees. They will never reach the same heights again in the conflict in political, moral and military terms. Whether they grasped the full import of the stage they had reached in their struggle is arguable but their military achievement was certainly not lost on them. "Never before have such modern weapons and equipment been used in this region" bragged the LTTE from Madras after the attack on the Jaffna police station. "Never before in the history of the militant struggle have such explosions been heard."[66] A diplomatic source was more sober but no less perceptive in his assessment: "the days of (bolt action) Lee Enfield rifles are over," he observed wryly.[67]

---

[66] 'Tigers V-sign', *Saturday Review*, 20.4.85, p.8.
[67] 'A New Spiral of Violence'.

2.5 The East

The Tigers' raid into Anuradhapura in April 1985 demonstrated that the militants were willing to take the fight to the South and were able to do so with impunity. The massacre sent shock waves through the Sinhalese community who felt that the war was threatening to spill over into Sinhalese dominated areas, spiralling out of control. However, contrary to such fears, the south was spared the violence, at least for the moment. The East, however, was not so lucky.

The Eastern Province was an integral part of the territory the militants were fighting for, part of the "Traditional Homeland" of the Tamils. But the troubles started slowly there, the East remaining relatively calm well after 1983 riots while the north simmered and boiled. A jail break on 27th September, 1983 saw a large number of suspected militants escape from the Batticaloa prison and in May 1984 a police informant was shot dead in Kalkudah.[68] But apart from such rumblings there were no significant incidents. This changed with the intensification of violence in August 1984, the war quickly spreading to areas outside the Jaffna Peninsula. As already mentioned, a landmine killed six soldiers in Mannar on 11th August. Soon, the East too was rocked by the sound of landmines and gunfire that took a steady toll in lives.

However, launching the struggle for Tamil Ealam in the East was different from waging the separatist war in the north. The Eastern province is a far more complex region than the Jaffna peninsula and the northern mainland. It is geographically extensive with a long coastline. Its eastern boundary is the coast that extends all the way south from

---

[68]'The Pirapaharan Phenomenon', chapter.15,
http://www.sangam.org/articles/view2/?uid=539

the southern banks of the Kokkali lagoon in the north to the eastern end of the Hambantota district. It is relatively narrow, embracing the immediate fertile hinterland of this coastal belt, except near Batticaloa where it spills deeper inland up to the central foothills. The province is divided into three districts, Trincomalee, Batticaloa and Ampara.

In some ways, the East was favourable to the guerrillas. Its geographical features, particularly the extensive coastline with numerous inlets and harbours, made it easier for the guerrillas to feed men and material into the province. The area northwest of Trincomalee harbour and the mainland to the west of the Batticaloa lagoon up to the border with Polonnaruwa, was also thickly forested and sparsely populated. Western Batticaloa in particular provided an ideal sanctuary for guerrilla activity with vast forested tracts, dominated by the 712 foot high Thoppigala rock and populated mainly by Tamils. The province also had a long border with Sinhalese dominated districts of Anuradhapura, Polonnaruwa and Hambantota which made them vulnerable to guerrilla infiltrations and attacks. Through Badulla, which had a large number of ethnic Tamil estate workers, inroads could be made into the hill country.

These factors were in favour of the guerrillas. However it was the demographic factor which made things complicated. The ethnic composition of the population of the Eastern province has undergone significant changes over the centuries. During the time of the Anuradhapura and Polonnaruwa kingdoms the eastern seaboard and its hinterland was ruled by Sinhalese kings with a predominantly Sinhalese population. However, as explained earlier, with the fall of Anuradhapura in the 11th century and the retreat of the Sinhalese kingdoms to the south in the 13th century, the eastern seaboard had come to be gradually settled by Tamil speaking people. By the turn of the nineteenth century the Tamilisation of the East was more or less complete. The 1901 census shows Tamils making up 55.83%

of the population of the province while Muslim form 35.97%. Sinhalese come a distant third with only 5.06%. However, by 1963 the Sinhalese population had risen to 19.88% with the Tamils making 45.03 and the Muslims 33.75%. By 1981 nearly a quarter of the population was Sinhalese.[69]

The influx of Sinhalese in the 20th century was the result of several irrigation schemes undertaken by the Sri Lankan state after Independence. The Gal Oya Scheme in 1956 dammed the Gal Oya river and opened up around 40,000 hectares in the eastern province for cultivation. It also brought in nearly 20,000 new colonists, the majority of them Sinhalese, to settle inside the western borders of the Batticaloa District that changed the demographic proportions. Two other irrigation projects, the Allai-Kantalai Project and the Padaviya scheme in the 1950s and the 1960s and the Gomarankadawela –Moraweva Projects in the 1970s and 1980s also resulted in the establishment of Sinhalese settlements, this time in the Trincomalee District. They were settled in lands in the north and south of Trincomalee District forming corridors of Sinhalese villages encompassing the Kottiyar Bay. As a result of these settlements, the percentage of Sinhalese settlers increased substantially. From a mere 4.2% in 1946, the Sinhalese population in Trincomalee had risen to nearly a third of the inhabitants in the district by 1981.[70]

Such settlement schemes increased the ratio of Sinhalese settlers to Tamils and Muslims, altering the ethnic balance in the province. In 1958 the government carved out a third

---

[69] *Sri Lanka's Eastern Province: Land, Development, Conflict*, Crisis Group Asia Report N°159, 15 October 2008, Appendix C p. 36
[70] 'Colonisation and Demographic Changes in the Trincomalee District and its Effects on the Tamil Speaking People,' University Teachers for Human Rights (Jaffna), Report 11, Appendix 2, http://www.uthr.org/Reports/Report11/appendix2.htm

district in the east out of the Muslim majority areas in the south and the newly colonised western part to form the Ampara District. In Trincomalee too, a new AGA division of Seruvila was created out of settlements where the vast majority of the inhabitants were Sinhalese settlers from outside the province.[71]

It is unclear whether these colonisation schemes were originally strategic in aim. The alteration of the demography of the province and the positioning of the settlements seem to suggest so. However, what is more clear is that the outbreak of the insurgency and the militants' stated aim of incorporating the east into their independent state of Tamil Ealam, the government found the Sinhalese settlers to be a valuable tool in undermining separatist strategy. Now, Sinhala colonisation became clearly strategic in aim, the insertion of more Sinhalese settlement between the Eastern and northern provinces calculated to break the contiguity of Tamil settlements in the Eastern and northern provinces. Accordingly, a string of new Sinhalese settlements along the *Maanal Aru* (Or Weli Oya as it was known in Sinhalese) was initiated, linking Padaviya with the Kokilai lagoon and the East coast. The Tamil villagers in these areas were gradually evicted after July 1983 in favour of the projected Sinhalese settlements. Just before Christmas 1984 the army asked several villages around Kokilai to be vacated within 24 hours.[72] Around the same time, another settlement project was planned along the Maduru Oya in the Batticaloa District.

The population in the Batticaloa District was largely Muslim and Tamil. The coastal belt between the 40 kilometre long lagoon and the sea, was heterogeneously Tamil and Muslim

---

[71] Ibid.
[72] 'From Maanal Aru to Weli Oya and the Spirit of July 1983', University Teachers for Human Rights (Jaffna), Special Report 5, http://www.uthr.org/SpecialReports/spreport5.htm#_Toc512569422

while the large expanse on the mainland to the west across the lagoon was Tamil dominated.[73] The Batticaloa town, a thin strip of land 2-4 kilometres long, was located on the coastal belt and also boasted a mixed Tamil and Muslim demography.[74] If successfully carried out, the Maduru Oya Project would have inserted Sinhalese settlers between the Batticaloa and Trincomalee districts, breaking the contiguity of the "Tamil Homeland" even further. However, the project fell through when it received wide publicity. It was planned as a secret operation by the Mahaveli Ministry to set up settlements of landless Sinhalese on state land. But when thousands of landless peasants converged on the Maduru Oya the scheme received unwanted publicity and when the few Tamil ministers in the government including the powerful S. Thondaman vehemently protested, the government found itself in a tight corner. In the aftermath of July riots, large scale colonisation of Tamil dominated areas was a very sensitive issue with India keeping a close eye on developments. The government backtracked, going so far as to arrest some of those responsible and removing the settlers. However, later, many of these settlers were found homes in the Weli Oya region.[75]

The establishment of this Sinhala 'buffer zone' – and the presence of other Sinhala colonists - set the tone for the separatist struggle and the fight against it in the eastern province. The Sinhalese settlers were an element missing from the war in the north. To

---

[73] D. B. S. Jeyaraj, 'LTTE Ascendant in the East', *The Island*, '9. 3. 97, p. 13.

[74] P. A. Ghosh, *Ethnic conflict in Sri Lanka and role of the Indian Peace Keeping Force*, http://books.google.com.au/books?id=YZscr75ijq8C&pg=PA131&lpg=PA131&dq=operation+checkmate+IPKF&source=bl&ots=0XZJyh-Run&sig=LSoPlhN80q_ayOqT_JmQF2lFVnc&hl=en&sa=X&ei=DspT4S4H8uuiQfcvuWtAw&ved=0CFUQ6AEwAw#v=onepage&q=operation%20checkmate%20IPKF&f=false, p. 129.

[75] See Guneratne, *For a Sovereign State*, pp. 232 - 4, 'The Pirapaharan Phenomenon', chapter 23, http://www.sangam.org/articles/view2/?uid=633

the Tamil militants, the Sinhalese settlers were clearly the allies of the state, planted to frustrate their drive for independence. Therefore in addition to attacking the security forces, a large part of the militants' strategy in the East involved the removal of these Sinhalese settlers. The execution of the strategy was ruthless and savage.

The first to taste the militants' wrath were settlers in Dollar and Kent farms in Mullaitivu. These were originally two Tamil owned business concerns. After the anti-Tamil riots in 1977 the two farms had been used to settle Tamil refugees of Indian origin from the plantations. Later the two farms were bought by the Prisons Department and became a cornerstone of the government's policy of Sinhalese settlement in the area. The Tamil settlers in the farms were evicted in mid 1984 and several hundred Sinhalese ex-convicts were settled in them. It was alleged that the settlers became a source of much harassment to the Tamil villagers in the area, acting with impunity due to the backing of the army.[76]

During the small hours of the 30th November 1984 two busloads of heavily armed LTTE Guerrillas approached the two farms. Led by Gopalaswamy Mahendrarajah or 'Mahaththaya,' one of Prabhakaran's trusted lieutenants, the Tigers crept in to the farms taking the inhabitants by surprise. They shot and hacked all the males they could get their hands on. Some of the victims at Dollar Farm were shoved into a room and blasted with explosives. Altogether 62 including three jail guards were killed at Dollar Farm and 20 more at Kent Farm.[77] The butchery continued the following day when another group of Tigers descended on

---

[76] 'From Maanal Aru to Weli Oya'.
[77] '84 Reported Slain as Guerrillas Raid Sri Lanka Farms', *Toronto Star*, 1.12.84

Nayaru and Kokilai, two villages of migrant Sinhala fishermen on the east coast south of Mullaitivu. When they finished, 59 fishermen lay dead, many of them killed in the most brutal fashion.[78]

The militants did not venture to attack civilians in the east for the next few months but by mid 1985 they were back in action. The Sinhala settlers encroaching into the 'Tamil homeland' now became a favourite target due to their strategic significance. On 30th May 1985 they killed five Sinhalese settlers at Dehiwatte in the Trincomalee District. A further eight Sinhalese civilians were killed on 12th June in Dehiwatte where more than forty houses were also burnt.[79] Security installations also continued to be attacked. On June 3rd the police station at Kuchchaveli was attacked by a large force of guerrillas. The 60-strong naval detachment at the police station held out against repeated attacks with rockets, grenades and small arms losing a sub-lieutenant. The guerrillas also blew up the bridge linking Kuchchaveli with Nilaveli.[80]

By mid 1985 Trincomalee was ringed by a number of guerrilla camps even extending in to the neighbouring Polonnaruwa district. The machinery for clearing the jungle was obtained by raiding nearby farms and government agencies.[81] There were attacks even within the Polonnaruwa District. For instance in mid-May 1985, the police guard room near the Manampitiya bridge was attacked leading to the death of two policemen.[82] To

---

[78] 'Pirapaharan Phenomenon', chapter 23, http://www.sangam.org/articles/view2/?uid=633, "Tamil Guerrillas kill 57 Villagers", *Canberra Times*, 3. 12. 84.

[79] 'Tamils Hit at Village', *The Age*, 13. 6. 85.

[80] '50 killed in Tamil Raids', *Sun*, 3. 6. 85. *The Sri Lanka Navy: a Pictorial History of the Navy in Sri Lanka 1937-1998*, Sri Lanka Navy, 1998, p.137

[81] Norman Palihawadana, 'Terrorists Get Away With Jeeps From Mahaveli and Maduru Oya Schemes', *The Island* 19.7.85.

[82] 'Manampitiya Police Guard Room Attacked-2 PCs killed', *The Island*, 21. 5. 85.

the north of Trincomalee the guerrillas controlled a long stretch of the coastline that extended up to Kokilai and its immediate hinterland. The blowing up of the bridge that connected Nilaveli to Trincomalee limited the military's access to the area by land, forcing them to rely more on the navy. The control of the coast in this part of the island was of great strategic advantage to the guerrillas; it had many inlets which became safe havens for the rebel boats which plied men and material between the north and the east.

As the war in the north intensified so did the bloodshed in the East. As in the north, the landmine became a favourite weapon of the guerrillas. In the aftermath of the second attack on Dehiwatte in June 1985, the guerrillas exploded a mine under an army convoy carrying a magistrate and a doctor to investigate the massacre, killing four soldiers and wounding two policemen[83]. According to a naval officer in Trincomalee all but one of the 40 military deaths that had occurred by September 1985 had been caused by landmines.[84]

Much of the guerrilla activity was in the countryside, away from the main urban centres but occasionally the militants also went for bigger targets in towns. In September 1985, a large group of guerrillas estimated at well over a hundred, attacked the police station at Eravur, killing seven constables. The police station, manned by 45 policemen, some of them raw recruits, held out against the sustained attack which lasted several hours. The attackers, who also included several women, had mined all approaches to the station and had caused a power blackout before the attack commenced. Several RPGs were fired at the police complex razing many of the buildings to the ground. The attackers had also blown up the culvert on

---

[83] 'Tamils Hit at Village'.
[84] Shekar Gupta, 'Terror Tactics', *India Today*, 15. 10. 85, p. 53.

Batticaloa- Eravur road to prevent reinforcements from arriving. Reinforcements from Batticaloa had to trek six miles, repulsing as many as three attacks to reach the station.[85] However, such attacks were the exception rather than the norm.

---

[85] 'Terrorists Kill Two TULF ex-MPs and Seven Policemen', *Ceylon Daily News*, 4. 9. 85, W. G. Gooneratne, 'Eravur OIC Bluffed the Terrorists', *Ceylon Daily News*, 7. 9. 85.

CHAPTER 3

Struggling to Respond – the Sri Lankan Military 1983-1985

The rebel onslaught posed serious challenges to the Sri Lankan state's nascent military apparatus. The armed forces were being forced to fight a war they were not trained to handle by an enemy whose ethnic composition differed from that of the majority of the Sri Lankan military personnel, especially those in the army. The Sri Lankan government's military response to the worsening rebellion in the north was largely shaped by these two factors.

3.1 A Different Enemy

The outbreak of the Tamil insurgency placed the government in a difficult situation. The new insurgency was very different to the JVP uprising. The JVP upsurge was sudden, widespread and more open in its methods of confrontation. The Tamil militancy, in contrast, was regional, ethnic in origin and its tactics clandestine. It was also part of a protracted campaign against the state. Whilst the JVP rebellion could be defeated by deploying superior resources, particularly firepower, the Tamil militancy needed to be handled with greater finesse.

The government first treated the militancy as a law and order issue. It was the police that was involved in hunting down the militants. But as the violence continued unabated, claiming even the feared Bastiampillai, the government became increasingly panicky. When Police Inspector

Guruswamy was shot down in Jaffna on July 1st, the government imposed a State of Emergency on the North.[1]

A brigade of 1500 soldiers was now dispatched to Jaffna under Brigadier Tissa Weeratunga. The instructions issued by the president J. R. Jayawardena to Brigadier Tissa Weeratunga makes interesting reading:

> 'It will be your duty to eliminate in accordance with the laws of the land the menace of terrorism in all its forms from the island and more especially from the Jaffna District. I will place at your disposal all resources of the State. I earnestly request all law abiding citizens to give their co-operation to you. This task has to be performed by you and completed before the 31st December 1979.'[2]

The instructions underlined the government's growing perception of the problem as one of simple terrorism or lawlessness that can be completely eradicated by force within a short period. The tone was regal – the description of the mission read more like directives given by a king to a general embarking on a campaign of conquest than that of a president entrusting a brigadier with the task of crushing an incipient rebellion by a few dozen ill-armed militants. It betrayed the arrogance of the rulers who saw the militancy more as a threat to the authority of the state rather than as an expression of simmering discontent. It was an arrogance that was to cost them and the country a great deal of blood and tears in the future.

---

[1] *Emergency '79*, Pamphlet published by the Movement for Inter Religious Justice and Equality, Kandy: 1980, p. 21.
[2] Quoted in S Sivanayagam, *Sri Lanka: Witness to History, a Journalist's Memoirs 1930 - 2004*, London: Sivayogam, 2005, p. 186.

However, the deployment of the army and spelling out of the mission also showed that even though the government was still treating the problem merely as one of law and order it was also taking the matter seriously in that context, like a real military challenge. If you want a fight, Jayawardena, now the first executive President of Sri Lanka, seems to be saying, I will give you real soldiers to fight against.

And the Brigadier took his job very seriously indeed. Sandhurst trained, he had already impressed with his service in Jaffna in the 1960s where he took part in operations against illegal immigrants, and then as the co-ordinating officer in Moneragala District during the JVP insurrection in 1971. Now he was determined to make his mark on a situation deemed to be steadily running out of control. The Army set up its headquarters in the official residence of the Government Agent of Jaffna. From there Weeratunga's men fanned out, determined to stamp out "the menace of terrorism." However, whether they carried out their duties "in accordance with the laws of the land" is arguable. Within a few days of the emergency coming in to effect, dozens of youth went missing in Jaffna, allegedly picked up by military men in the night. Several turned up dead bearing marks of torture while others remained missing. There were mass arrests and reports of public humiliations meted out to residents by the security forces who were behaving very much like an army of occupation.[3]

The crackdown, though brutal, was effective. The ruthless operations of the security forces resulted in a marked decrease in the militant activities. Many suspected militants

---

[3] *Emergency '79*, pp. 23-35.

were tracked down and arrested.[4] Fearing capture many of the leaders and the top cadres went into hiding, some overseas in Tamil Nadu. By the end of the year Weeratunga was able to proudly report to the president that he had accomplished his mission. A lavish party at the Rock House camp in Mutwal on Christmas Eve celebrated the conquering hero's return.[5] All was well with the world it seemed.

The lull in militant activities continued for more than a year. This, however, was a false dawn for the government. Although many militants were captured and the population had been terrorised the militancy had not been crushed. The insurgency of Jaffna was more complex than the one in the south. The JVP had no refuge across the seas but the Tamil militants did. The Tamil militant leaders escaped to India and while they were away they utilised their time to train themselves. The LTTE set up a camp in the Trichy district in Tamil Nadu with the connivance of local authorities to train its cadres.[6] Here they learned to fire their guns and bided their time until it was safe to return to Sri Lanka. Other groups did the same, lying low and waiting for the storm in Jaffna to blow over.

Thus, the Emergency did not crush the rebellion. It only suppressed it for a while. Moreover, by treating the people with a heavy hand the security forces had won few hearts and won many enemies. This created fertile ground for the rejuvenation of the rebellion when the militants returned from India. And they were back within a year. In late 1980, PLOT announced the return to violence by killing a supporter of the ruling United

---

[4] L. M. H. Mendis, *Assignment Peace*, In *the Name of the Motherland*, Nugegoda, Sri Lanka: Author Publication, 2009, p. 16.
[5] Munasinghe, *A Soldier's Version*, Colombo: author publication, 2000, p. 51.
[6] M. R. Narayan Swamy, *Inside an Elusive Mind*, *Prabhakaran*, Delhi: Konark Publishers, 2003, p. 57.

National Party, R. Balasunderam in Kilinochchi.[7] Soon, assassinations and robberies recommenced, the militants taking on police stations and even the army.

As the violence escalated the government realised its mistake of celebrating the crushing of the militants too early. The government had recalled the army after the insurgency had subsided in 1979-80, maintaining only a token presence in the North. Now, when the militants returned to the fray in 1981, the army too returned. Tissa Weeratuga, now a Major-General and the Commander of the Army, was sent back north to complete the job which now seemed only half finished. An army brigade was now to be stationed permanently in Jaffna.[8] But this time they found an enemy somewhat different to the one they had scared off barely two years ago.

To be sure, the army was far better equipped than the militants who were armed with just a few pistols, shot guns and grenades. The real problem for the army, however, was that it had never been prepared for the kind of war it was now called upon to fight. No matter how irregularly equipped and indifferently trained, and no matter what its leaders have been saying about preparing for internal policing, the Sri Lanka army was geared to fight a more conventional enemy than the bomb-throwing and sniping Tamil militants. Even the JVP insurgents, although deadlier than the early Tamil militants, with all their amateurish enthusiasm, had presented a more tangible enemy. They often came to attack in large groups and tried to capture territory and installations often using frontal attacks. The Tamil Militants were different; they operated in small

---

[7] Mendis, *Assignment Peace*, p. 21.
[8] T. D. S. A. Dissanayake, *War or Peace in Sri Lanka*, Colombo: Popular Prakashan, 1995, vol. 2, p. 28.

groups, sometimes even individually or in pairs, and relied on stealth, surprise and speed to achieve their objectives which were killing of security personnel and informants and the capture of weapons. They blended easily with the public. They had no intention of taking over towns and establishing their authority. At least not yet.

This is not to say that the army suffered horribly at the hands of the militants at this stage. Far from it. From 1981 till July 1983 far fewer soldiers died in the north than during the few weeks of the JVP insurgency.[9] What was ominous was the clandestine nature of the attacks which took the army by surprise portending serious challenges ahead.

In the sea the challenge was no less difficult. The tiny Sri Lankan navy possessed only a handful of patrol crafts with which to prevent the guerrillas from smuggling arms and men. The military commander in Jaffna Brigadier Nalin Seneviratne lamented that he had about 150 miles of coastline to protect and that if access from nearby countries could be sealed off the battle would be 80% won.[10] But this was only wishful thinking as long as the Sri Lankan navy consisted of only a motley collection of patrol boats and aging gunboats. The Air Force was of not much use either; the few Bell 206 helicopters were helpful, but not sufficient to cover the long stretches of coast.

---

[9] According to A. C. Alles, 37 policemen, 19 Army personnel, four airmen and three sailors were killed during the period of the insurrection which occurred during the month of April 1971. Total of 130 members of all security forces were wounded. A. C. Alles, *The J.V.P. 1969 – 1989*, Colombo: Lake House, 1990, Appendix I, p. ii.

[10] 'Now, the Brigadier Speaks'. *Saturday Review*, 25.8.84, p. 12.

3.2 Meeting the Challenge

As the situation in the north began to deteriorate the government began to look for ways to counter the challenge. In March 1983 Lalith Athulathmudali was appointed the Minister for National Security and in the same month Brigadier General Nalin Seneviratne was placed in command of the troops in Jaffna.[11] Seneviratne, a tough no-nonsense officer was seen as the proper person to lead the drive against the militancy.[12] As explained earlier the army's presence was also augmented. Immediately after the killing of the two soldiers in October 1981 the existing army detachments at Velvetithurai, Madagal and Elephant Pass had been brought up to the strength of a company in each place[13]. When the Tamil "offensive' began in August 1984 the number of camps were also increased. By the end of the year there were six army camps in the peninsula while more were planned for the new year.[14] Elements of armour were also deployed in Jaffna and also in camps in the Vanni.[15] At the same time recruitment was also stepped up. The army had stood at just over 11,000 in 1983 but with the outbreak of sustained violence in August 1984 the army embarked on a rapid expansion. The response to the army's call for recruits was enthusiastic, thousands of young men "barely out of their teens" queuing up outside the army headquarters to enlist.[16] In the early 1980s two new regiments, Rajarata Rifles and Vijayaba Regiment were also formed. These were amalgamated in 1983 to form the Gajaba Regiment. The plan at

---

[11]'Army Officer for Jaffna', *Malaysian Straits Times*, 29. 2. 84.

[12] Shekar Gupta, 'Haven in India for Lankan Guerrillas', *Sydney Morning Herald*, 14.4.84.

[13] *SL Army 50 Years On*, Colombo: Sri Lanka Army, 1999, p. 347

[14] Suvendrinie Suguro, 'More Army Camps Planned', *Daily Observer*, 31. 12. 84.

[15] Jagath P. Senaratne, *Sri Lanka Armoured Corps: 60 Years of History* 1955 - 2015, Sri Lanka Armoured Corps, 2015, p. 53.

[16] 'Thousands Answer Army's Call', *Weekend*, 9.12.84, p. 1.

this stage was to increase the numbers by 2000 every two months. The training period was cut down to 8 weeks from three months to speed up the process.[17] By 1985 the army had expanded to 16,000 and 30,000 in 1986. By 1987 it had grown further, standing at 40,000 men.[18]

The police also took measures to meet the growing challenge of an incipient guerrilla movement. In 1983 a Special Task Force (STF) was formed to provide a combative edge to the police. The STF was a paramilitary police force to boost the Sri Lanka police in its operations against the militants and consisted of a small force of picked men trained by the army in the handling of military weapons. In February 1985 the government also created the Joint Operations Command (JOC) to co-ordinate actions by all three armed forces. Major-General Cyril Ranatunge was brought back from retirement to be its GOC.[19]

In the sea the government tried to overcome the challenge of inadequate resources by enforcing a harsh measure. In April 1984 a naval surveillance zone was also declared around Jaffna, where boats with outboard motors were banned. In November the same year this zone was tightened by the declaration of a "prohibited zone' that extended 100 meters into the sea. Entry to it was only with permission of the local police station. In November 1984 the government also banned all fishing off the northern coast.[20] Similar

---

[17] Iqbal Athas, 'Sri Lanka Strengthens Defence Forces', *Jane's Defence Weekly*, 3, 2, (12.1.85) p. 45.

[18] Brian Blodgett, *Sri Lanka's Military: the Search for a Mission*, San Diego, California: Aventine Press, 2004, p. 93.

[19] Cyril Ranatunge, *Adventurous Journey*, Colombo: Vijitha Yapa, 2009p. 109.

[20] Edgar O'Ballance, *Cyanide War: the Tamil Insurrection in Sri Lanka 1973-88,* London: Brassey's, 1989, p. 36, Thomas A. Marks, 'Insurgency and Counterinsurgency', *Issues and Studies*, August 1986, pp. 82-3.

measures were also taken on land; after the spike in violence in August 1984 the military banned the use of private vehicles in Jaffna. They also launched large scale cordon and search operations that were aimed at netting militants and their sympathisers. One of the earliest and biggest such operations took place between 9 and 12 December 1984. Jaffna, Kilinochchi and Mullaitivu were placed under a 42 hour curfew and hundreds of soldiers supported by armoured vehicles ranged through the cities in focus detaining of over 700 youths and the capturing some weapons and documents.[21] According to a Tamil MP the army usually surrounded a village early in the morning and ordered men in the age group of 18-35 to report at a designated place. The houses were then searched and if any young men were found there they were 'severely dealt with.' Those who went to designated place were questioned and if deemed suspicious, sent to Boosa in southern Sri Lanka for further detention and interrogation. By January 1985 there were already close to 800 such detainees in Boosa.[22]

While squeezing the population in the north the government also sought to temper coercion with persuasion, but of a crude kind. They offered cash rewards for information on militants weapons. An SLR was worth Rs, 25,000 while a .303 rifle fetched Rs. 5000.[23] However, it is not known whether this enticed the public to divulge the desired information.

Facing a rejuvenated militancy, the government also looked to enhance the effectiveness of its growing military forces

---

[21] 'Operation Search and Destroy Launched', *Weekend*, 9. 12. 84. p.1,O'Ballance, *Cyanide War*, p. 43
[22] 'Siva Tells the Hindu – Jaffna Area a Prison', *The Sun*, 21. 1. 85.
[23]Mervyn De Silva and S. Venkatramani, 'Reign of Terror', *India Today*, December 31, 1984, p. 23.

with new equipment and specialised training. They needed to boost the army's defence against the threat of landmines and explore more effective ways of carrying out surveillance and interception in the sea. This however was not easy to achieve. Funds were a major problem. Sri Lanka in the early 1980s was no economic powerhouse and even though the economy was just beginning to forge ahead under the liberal policies introduced by the government in 1977, the escalating violence was threatening to undermine it. Sophisticated weapons and equipment were not quite within the means of such a struggling economy. A bigger problem was the lukewarm response from potential sources of help. In mid 1984 President Jayawardena visited China and the US seeking assistance but got little more than sympathy. Lalith Athulathmudali also visited the US in January 1985 but apart from a promise to share and exchange information on terrorism the US was not willing to provide much else.[24] One major reason for this was that the government's cause was not greatly appreciated by those who could help, especially in the aftermath of the riots in July 1983. The Sri Lankan military too did not seem in a fit state to receive aid. Vernon A. Walters the roving US ambassador is said to have even expressed the concern that the Sri Lankan army was on the verge of mutiny.[25] There was also the desire not to antagonise the regional power India which clearly had a stake in the outcome of the struggle in Sri Lanka. In the context of the Cold War, with India developing close relations with the Soviet Union, the USA was not averse to cultivating Sri Lanka as an ally in the Indian Ocean but the Americans were not willing to do that conspicuously.

---

[24] 'The Know-how to combat Terrorism', *Sun*, 26.1.85.
[25] O'Ballance, *Cyanide War*, p. 52.

American help came in a more subtle way. In May 1984 Sri Lanka announced that Israel has opened a "special Interest Section" in Sri Lanka. The move was facilitated by the US who allowed the Israeli interest section to function from the American embassy in Colombo. The Israelis were from the internal security agency known as *Shin Bet* and they were to help with counter-insurgency training for the security force.[26] The government denied that the foreigners were engaged in any combat but only employed as "consultants" in training the local security forces in intelligence gathering.[27] President Jayawardena explained the decision to turn to Israel as necessitated by the need for foreign help in a context where little help was forthcoming.[28] Expertise with counterinsurgency came from other sources as well. In 1984 the government obtained the services of a British Security Firm KMS services which employed former SAS personnel. The British government, despite its reluctance to provide direct military aid to Sri Lanka, turned a blind eye to KMS offering its personnel and expertise. Accordingly, a number of KMS personnel arrived in Sri Lanka to organise training for the Special Task Force. Initially consisting of only a handful of combat trained policemen, this force was now expanded under the guidance of foreign expertise and trained to handle riot situations, sophisticated weapons, counter-insurgency and fighting in built up areas (FIBUA).[29]

In the meantime, the search for new equipment continued. Undaunted by the reluctance of Western nations to help, the

---

[26] Ibid., p. 37.
[27] 'Israeli, British Agents Helping Lankan Forces', *New Sunday Times*, 12. 8. 84.
[28] 'Sri Lanka 'forced' to Seek Israeli Help', *The Straits Times*, 3. 7. 84.
[29] Phil Miler, *Britain's Dirty War Against the Tamil People, 1979 - 2009*, Bremen: International Human Rights Association, 2015, pp. 12-15, Thomas A. Marks, 'Sri Lanka's Special Forces', *Soldier of Fortune*, July 1988, p. 35.

government sought arms and equipment from other sources that were willing to supply them without publicity. These included Italy, Israel, Yugoslavia and South Korea.[30] Military expenditure began to rise, as the embattled government scraped its coffers, obtaining 2 million rupees (US $ 875,000) by March 1985. [31] The cost would continue to grow in the future, becoming a severe burden on the struggling economy.

Predictably the equipment purchased was of a modest nature, as befitting the financial means of the government. Immediately after the ambush in July 1983 several Shorland and Hotspur armoured personnel carriers were purchased to provide better protection against landmines and ambushes. Placed in a new armoured squadron under Major P. A. Karunatilleke, the APCs provided greater protection to the army's patrols in the North.[32] The army also began gradually arming its troops with the Chinese made T56 assault rifle and RPGs to match the growing sophistication of the enemy.

The air force also got new wings. When the war escalated in 1983 the Bell 47 Gs were not in service anymore and the Air Force had to depend on its small fleet of Bell jet rangers to carry out reconnaissance, transport and interception duties. Some of these were fitted with 70mm rocket pods on one side and .50 calibre machine gun pods on the other.[33] The government was also looking for a better

---

[30] O'Ballance, *Cyanide War*, pp. 52-3. KMS services has now rebranded itself as Saladin Security. Miler, *Britain's Dirty War*, p. 18.
[31] Denzil Peiris, 'Colombo Rides the Tiger', *Weekend*, 3. 3. 85, p. 6 and p. 22.
[32] Mendis, *Assignment Peace*, p. 44, Blodgett, *Sri Lanka's Military*, p. 92.
[33] Robert Craig Johnson, 'Tigers and Lions in Paradise: the Enduring Agony of the Civil War n Sri Lanka',

helicopter to be used in surveillance and attack but the lack of funds prevented them from obtaining dedicated gunships. They had to settle for the Bell 212. Two were purchased in 1984 and 12 more ordered. Nine of these were to be fitted with 12.7 inch machine gun pods and rocket pods. The first two arrived in Sri Lanka in February 1985 while four more were delivered in July that year.[34]

In the early to mid 1980s the Sri Lanka navy also acquired more vessels in order to police this coastal belt. These were several coastal patrol craft and a number of inshore patrol crafts which were produced by the Colombo Dockyards. In 1984 they also purchased 6 Israeli built Dvora class patrol boats armed with 20mm cannon. These were supplemented by the acquisition of ten Cougar Class inshore patrol boats.[35]

3.3. The Shifting Balance

Thus by mid 1980s, the security forces were gradually enhancing their defensive and offensive capabilities. However, these modest improvements did not immediately translate into a decisive advantage over the militants. In the sea, the addition of the Dvoras was a welcome boost to the navy's ability to intercept militant boats darting between Tamil Nadu and the north but they were not sufficient to stem the traffic significantly. The few armed helicopters had a similar effect. On land the APCs did afford a degree of protection, but mainly against small arms. This was

---

http://worldatwar.net/chandelle/v3/v3n3/articles/srilanka.html , Tom Cooper, 'Sri Lanka Since 1971',
http://www.acig.info/CMS/?option=com_content&task=view&id=174&Item id=1
[34] Peter Steinemann, 'The Sri Lanka Air Force', Asian Defence Journal, Feb. 1993, p. 58.
[35] Blodgett, Sri Lanka's Military, pp. 102-3.

indeed a blessing as, despite the improvement in their arsenal since July 1983, the guerrillas still possessed few RPGs which would have penetrated the APCs easily. However, the landmines, particularly the bigger Improvised Explosive Devices (IED) still ripped the armoured vehicles apart. Moreover, the APC's were all wheeled vehicles which gave very limited cross-country manoeuvrability making them more vulnerable to mines. Even the measure of protection provided by the APCs was available to only a fraction of the army as it was able to afford only a few APCs. The vast majority of the soldiers had to rely on standard military trucks and jeeps. Often the cash- strapped army even had to commandeer civilian buses and lorries to carry its soldiers and it was also not uncommon for military patrols to be out on foot. The result was continuing heavy casualties to the troops from landmines.

And these losses came at little risk to the guerrillas. Despite their growing numbers and more sophisticated arsenal, the troops suffered from hit and run attacks from an enemy who was too elusive to be blasted with their heavy firepower. There was hardly any 'fighting' – except when the militants took on a police station or a military outpost. This, however, was rare. What was common was the ambush with landmines. The first and perhaps the only indication of the presence of the guerrillas was the explosion. After that, the survivors and reinforcements were left to pick up the pieces, literally. While the guerrillas presented only a fleeting target to the army, striking stealthily and disappearing rapidly, the army was constantly offering them targets with their convoys and outposts. The struggle was certainly skewed in favour of the militants.

The only way the army could avoid this was by being proactive and taking the fight to the militants but that required stealth and information from civilians. Before the riots in 1983 this was forthcoming to some extent; a case in point is the information that led to the killing of Seelan, one of the key LTTE operatives. Acting on intelligence received from their informants, the army ambushed and killed in July 1983.[36] But after the July riots the army was increasingly operating in hostile country in the north. Most Tamils had come to see the military as an alien, occupying force and the few who were willing to offer information were brutally silenced by the militants. But even then information trickled in. In January 1985 the army launched a very successful raid on an LTTE hideout in the Achchuveli area of the Jaffna peninsula capturing a large haul of weapons, which consisted of two lorry loads of arms and ammunition including several RPGs and nearly 20,000 rounds of ammunition stored in an elaborately constructed bunker. The raid also killed 14 Tigers which included a prominent leader, Sinnadurai Ravindran or 'Pandithar.' The raid was the result of information received from the public. But such success was a rare luxury for the security forces as the civilians, either out of fear for the militants or loathing for the army, chose to remain silent.[37]

The situation was made worse by the fact that the army was utterly inexperienced in combat of any kind, let alone urban guerrilla warfare. Apart from a few officers who had served in 1971 few if any ordinary soldiers had seen any real action. Many of the soldiers who had to patrol the landmine infested streets of Jaffna in late 1984 and early 1985 were raw recruits, sent north after basic training which had been cut down to eight weeks. Needless to say they found the war they had to fight particularly intimidating. Even the basic training they received mattered very little in the north

---

[36] Munasinghe, *A Soldier's Version*, pp. 88-92.
[37] Ibid., p. 17, Roland Edirisinghe, 'Sri Lanka Links Rebels Base to India', *Sydney Morning Herald*, 12. 1. 85.

where the enemy was often invisible and the only indication of his presence was a landmine, a bomb or gunfire. It made the green troops feel utterly vulnerable without being able to hit back.

The result was the gradual adoption of a defensive approach. The police gave in first. As the pressure from the militants increased, police stations began to shut down. The process had begun way back in July 1981after the attack on the Annacottai police station. After the attack the government decided to close some of the smaller police stations and strengthen the bigger ones[38]. After the devastating attack on the Chavakachcheri police station all the outlying police stations were closed and the policemen concentrated on the two stations in Jaffna and Kankesanthurai. After the attack on the Jaffna police station in April 1985 police stations ceased to exist in Jaffna.

With that the army became the keepers of the government's writ in Jaffna. The army held on to their camps. The problem for the troops however, was projecting their power beyond the confines of their bases. The militants had set up outposts around the camps keeping a close watch on the soldiers. They even set up roadblocks and bunkers within meters of the army camps and mined all approaches to the camps.[39]

The soldiers learnt to deal with the problem by moving out of camps only if it was necessary and then only in strength. Gone were the days when an officer could venture out with only his pistol and feel safe. Now the smallest unit that

---

[38] Virginia A. Leary, *Ethnic Conflict and Violence in Sri Lanka*, report of a mission to Sri Lanka n July-August 1981 on behalf of the International Commission of Jurists, International Commission of Jurists, 1983, p. 27.

[39] S.Venkatramani, ' Bearing the Blockade ', *India Today*, 15.2.87, p. 25.

dared move out was a platoon – about 30 men.[40] As early as 1982 it was said jokingly that a whole truckload of soldiers was needed to buy a packet of cigarettes.[41] That was no doubt an exaggeration. But after the upsurge in violence in 1984, such absurdities became much closer to reality. During night time, even a truckload of soldiers dared not travel beyond the perimeters of the camps.

As a consequence, the towns and villages in the north were gradually being ceded to the militants. There was little if any patrolling in strength, a huge drawback in a war of this nature. The population for whose control the army was competing with the guerrillas had to see that the army was not afraid to venture into 'enemy territory.' It inspired respect for the army and also had the potential to boost the morale of those who might be willing to provide information. Instead, the army chose to stay cooped up in their camps. This enabled the militants to organise themselves with impunity and launch bigger attacks involving larger numbers. The attack on the Jaffna police station was a case in point.

The availability of air support alleviated the problem to an extent. In the air the Sri Lankan forces were relatively free from guerrilla interference and therefore were able to use their airpower to ferry troops and supplies to the increasingly beleaguered garrisons and to attack suspected guerrilla targets. This was a significant advantage against an enemy who possessed few heavy weapons and almost no conventional anti aircraft capability. It gave the air force considerable command of the air which became invaluable in the siege situation that was gradually developing in the north. The helicopters in particular played an important role in supplying outposts and

---

[40] ' Is there a Way Out?', *Asiaweek*, 21. 6. 85, p. 32.
[41] Sivanayangam, *Sri Lanka: Witness to History*, p. 215. The writer is quoting David Selbourne, 'Sinhalese Lions and Tamil Tigers', *Illustrated Weekly of India*, 17. 10. 1982.

providing air cover to ground troops. But still one must not overestimate the advantage offered by the few helicopters. They were not able to ferry troops in large numbers and their small number meant that their availability was limited in the face of escalating violence in the north. Having armed helicopters helped but the machine guns and rocket pods also encumbered the machines affecting their performance. The Jet Rangers suffered most from the additional fittings. It took nearly ten minutes for the helicopters to climb 1000 feet and the whole machine became unstable when firing the weapons and descended to 600 feet during the operation.[42] Even the Bell 212 became heavy and un-manoeuvrable with its full complement of machine guns and rockets; "like an overloaded shopping trolley it tended to go pretty much where it liked' according to one pilot.[43] They were also not totally invulnerable to rebel arms. In November 1984 a helicopter came under fire from what was thought to be a high-powered rifle while trying to land in Jaffna. The landing was made possible only after ground fire had silenced the enemy fire.[44] The helicopter that went to intercept the guerrillas responsible for the Anuradhapura massacre suffered heavily from the enemy's small arms fire despite inflicting some casualties on the guerrillas. Several commandos were wounded while the helicopter itself was badly damaged.[45] Sometimes the fledgling air force could be as vulnerable as the novice army.

---

[42] Steinemann, 'The Sri Lanka Air Force', pp. 56-8.

[43] Tim Smith, *Reluctant Mercenary: the Reflections of an Ex-Army Helicopter Pilot in the Anti-Terrorist War in Sri Lanka*, The Book Guild Ltd., Sussex 2002, p. 49.

[44] 'SLAF chopper Comes Under Sniper Fire'. *Weekend*, 25. 11. 84, p. 1.

[45] Shamindra Ferdinando, 'Armless veteran talks of War and Peace', *The Island*, 27. 10. 02, http://www.island.lk/2002/10/27/featur08.html
Rohan Gunaratna, *War and Peace in Sri Lanka*, Colombo: Institute of Fundamental Studies, 1987, p. 43.

With few ways of hitting back at the enemy the inexperienced and much harassed troops resorted to an option which was as predictable as it was brutal. They unleashed indiscriminate violence on civilians.

3.4 The War on Civilians

The police set the trend from an early stage. One of the earliest and most infamous police rampages occurred on the night of the 1st of June 1981 when a group of policemen ran amok through the streets of Jaffna after an attack on some of their colleagues. The trouble had started the previous day, the 31st May at Nachchimar Kovilady. The TULF was holding a meeting as part of its campaign in the District Development Council Elections. A group of militants had approached the four policemen detailed to provide security and opened fire. Two of the policemen fell dead while the others were wounded.

The attack was the trigger for two nights of violent reprisals by the police who roamed the city in vehicles, burning and shooting. They set fire to a kovil, the Jaffna MPs house and the TULF office in Jaffna. On the second night they landed their cruellest blow: they set fire to the Jaffna public library.

For decades the Jaffna Public Library had been the pride of place in the intellectual and cultural life of the peninsula. Starting as a private collection in 1933 it had grown into a major library by 1960, thanks to the contributions from various individuals. It contained tens of thousands of volumes of books and journals and also a priceless collection of *ola* manuscripts and private records. On the night of 1st June 1981 this icon of Jaffna culture went up in flames. According to some witnesses the fire began in the third floor where the rare manuscripts were housed and then quickly spread to the other floors. By the following morning it was all over. All that remained of the library was its hull brooding like a ghost over the remains of incinerated books and papers.[46]

The fire was no accident. It was a deliberate act of arson perpetrated by a gang of policemen on the rampage in Jaffna. It was also not the last such outrage by security personnel and was followed by many others, some of them chillingly horrific. When a soldier was killed and his assault rifle taken in May 1983 his colleagues took matters in to their own hands, burning scores of houses and vehicles.[47] When two airmen were killed in Vavuniya on 1st June the same year, airmen went on the rampage, burning and destroying. One of their victims was the Gandhiyam farm just 1.5 miles outside Vavuniya.[48] All these attacks were still directed more against property than people. That changed dramatically with the ambush in Tinneveli in July 1983. Sri Lankan soldiers from the Madagal, Palali and Velvetithurai camps broke out of the camps and went on the rampage, shooting civilians. Some officers struggled to maintain discipline but to no avail.[49]

As the violence escalated in 1984 so did the civilian toll. When an Army truck was attacked with a car bomb in April 1984 the soldiers hit back with violence. According to the Government Agent of Jaffna, Devanesan Nesiah the army shot dead about 50 people during the rampage.[50] After the landmine attack in Velvetithurai on 5th August 1984 the forces went berserk destroying houses and businesses, shooting people indiscriminately. A terrified Jaffna

---

[46] Sivanayagam, *Sri Lanka: Witness to History,* pp. 193-8.
[47] Rajan Hoole et al, *Broken Palmyrah, The Tamil Crisis in Sri Lanka, an Inside Account,* Ratmalana, Sri Lanka: Sri Lanka Studies Institute, 1992, p. 57
[48] Ibid., p. 59.
[49] Munasinghe, *A Soldier's Version,* pp. 8-9.
[50] 'More Than 50 Killed in Tamil Violence', *Canberra Times,* 20.4.84.

resident claimed that it was "as if the soldiers had been given permission to fire at will' and that they were shooting at "anything that moves".[51] According to the Velvetithurai Citizens Committee sporadic rampages between 4 and 9 August destroyed 130 houses, 90 shops, four cars and six fishing boats while 29 boats were also confiscated by the Navy.[52] Military detachments passing through the hostile countryside often left scenes of death and destruction in their wake. A group of foreign journalists driving through Achchuveli in Jaffna in mid August 1984 met a convoy of armoured vehicles accompanying troops on foot, firing into deserted houses at random and setting houses on fire.[53] When a landmine killed six soldiers in Mannar on the 11th of the same month the troops ran amok burning more than 100 shops. A few months later, Mannar suffered again, this time in response to a landmine attack that killed a soldier and wounded 6 others near Jubilee Junction on the Mannar-Medawachchiya road. According to the Mannar District Citizens Committee the army went berserk shooting civilians in homes and vehicles. When the soldiers ordered the passengers in one Ceylon Transport Bus to alight the conductor, a Sinhalese protested, saying he was responsible for the lives of the passengers and that he had to be killed before the passengers were harmed. The soldiers shot him dead before shooting the passengers. Altogether, 90 people were reportedly killed that day.[54]

---

[51] William Clairborne, 'Tamils Hide in Fear as Troops take Revenge', *The Age*, 15. 8. 84.

[52] 'Velvetithurai, A Fishing Village Victim of Pogrom', *Financial Times*, 21. 8. 84, reproduced in S. Sivanayagam (ed), *Tamil Information*,(published for private circulation), Madras 1. 9. 84, vol. 1, nos. 4 and 5, p. 12.

[53] Sayeed Naqvi, 'How the BBC Man Faced Jayawardena's Cannon', in S. Sivanayagam (ed) *Tamil Information* (published for Private circulation), Madras, 1. 9. 84, vol. 1, nos. 4 and 5, (The article originally appeared in the *Indian Express*). p. 17.

[54] 'Mannar Tragedy', *Saturday Review*, 22. 12. 84, p. 3.

The war was steadily acquiring a murderous pattern: the militants setting off a landmine and disappearing and the military retaliating by targeting civilians. But murder and arson were not the only components of the Military's rampages. A young woman, eight months pregnant, described to a foreign journalist in December 1984 how a soldier raped her at gunpoint while his colleagues set fire to her house. The journalist had first-hand experience of the terror inspired by such brutality when he witnessed the civilian reaction to an army patrol stopping in Jaffna town. As soon as the patrol stopped in the middle of the town, dozens of men, women and children, ran away in terror.[55]

Coupled with the cordon and search operations, these bouts of violence spread panic among the people in the North. Many people fled to India to escape the mayhem as the terror continued well into 1985. Velvetithurai was singled out for more punishment in May 1985. After a landmine attack killed several soldiers in the first week of May, the soldiers ran berserk in the town killing and burning. Velvetithurai Citizens' Committee supplied the names of 40 killed, many of them under the age of 25. It was alleged that 24 of these youths were herded into a community centre and the building blasted.[56] The Tigers' raid on Anuradhapura four days later, it was also alleged, was in retaliation for this rampage.[57]

Caught in the crossfire between the militants and the security forces Jaffna soon turned into a ghost town. Its roads came to be lined with burnt down buildings and pock-marked with dozens of craters caused by landmines. Shops

---

[55] David Graves, 'Troops Tackling Rebels in Divided Sri Lanka Terrorise Tamils', *Daily Telegraph*, 17. 12. 84.
[56] 'Blood, More Blood, on Our Hands', *Saturday Review*, 18. 5. 85, p. 1, 'V.V.T. Massacre', *Saturday Review*, 25. 5. 85, pp. 5, 7 and 8.
[57] Vilma Wimaladasa, '86 Killed in Tamil Raid on Holy City', *Daily Telegraph*, 14. 5. 85.

pulled down their shutters by 4 PM.[58]A western journalist visiting the region in January 1985 was struck by the number of burnt out shops and houses she encountered during a drive in to the peninsula. In eight hours of driving her vehicle was the only private vehicle on the road.[59] At night an eerie quiet descended on most parts of the peninsula as only the armed militants roamed the streets.

In the East too, the civilians suffered at the hands of the security forces. This was most apparent in the 'border villages' which were becoming a battle ground between the two warring sides. When the Tamil militants attacked Dehiwatte in June 1985 retribution came in the form of another bout of violence against Tamil civilians in the area. The security forces joined Sinhalese mobs in burning down hundreds of houses in adjacent Tamil villages.[60] One villager from Thiriyay described to a foreign reporter what happened when the army arrived in June 1985:

> "A helicopter came first at 8 AM firing guns down at us. Then the lorries came with hundreds of soldiers. They fired their guns and drove us all out into the jungle. They poured paraffin on the houses and burnt them. They went very quickly. They were gone by 10 am. By then the whole village was on fire."[61]

---

[58] Trevor Fishlock, 'Economy Drained as Tamils Flee City', *The Australian*, 4. 1. 85.

[59] Maryanne Weaver, 'Civil War Looms With Separatists', *The Australian*, 31. 1. 85.

[60] 80 Tamils die in Lanka Raids', *Herald*, 6. 6. 85, 'Tamils flee as villages are burnt and looted', *The Age*, 12. 6. 85.

[61] Simon Winchester, 'Behind the Lines with the Tamil Guerrillas', *The Australian*, 29. 6. 85.

Around the same time, a similar fate befell the village of Kuchchaveli. According to a villager, the soldiers arrived in armoured cars firing wildly. Then a helicopter appeared and began firing, killing and injuring several villagers. He and his nine year old son escaped by hiding in the jungle.[62]

In Batticaloa the STF was often accused of torture. "We can't go on the road without being threatened. We can't go to our shops, we have no freedom at all" a man complained to a foreign journalist.[63] Sometimes reprisals went beyond torture. In May 1985 at least 40 Tamil civilians in Kalmunai in the Ampara District were gunned down, allegedly after being forced to dig their own graves. The police confirmed the killings by "unidentified attackers" but the massacre was widely attributed to the STF.[64] The police commandos were often blamed for brutal, often casual murders of civilians, sometimes as casual as taking pot shots at peasants from passing vehicles.[65]

The number of civilians killed in these terrible bouts of retribution is hard to estimate. The *Saturday Review*, relying largely on Citizens' Committees in the north who kept meticulous records on the details of the dead, claimed that 2,215 civilians were killed during 1985.[66] This was probably close to the truth. One unnamed government official went on to claim that the maximum number of people killed in a single attack since 1981 was around 75 or

---

[62] Humphrey Hawksley, 'How the Battle for Trinco Was Won', *Lanka Guardian*, 1. 7. 1986, p. 13.

[63] Steven Weisman, 'Terror on the Beach in Sri Lanka', *Sydney Morning Herald*, 12.2.85.

[64] 'Latest anti-Tamil violence claims 56', *The Australian*, 20. 5. 85, '59 Tamils murdered', *Sun* 21. 5. 85.

[65] Hoole et al., *Broken Palmyrah*, p. 119.

[66] 'Bleeding Statistics', *Saturday Review*, 4. 1. 86, p. 4.

100.[67] Violence on civilians became so commonplace that Western diplomats described Sri Lanka's tiny army as one of the most undisciplined in the world.[68] It was such observations that prompted General V. Walters, President Reagan's special envoy to go so far as to warn President Jayawardena that his armed forces were "running out of control."[69]

Given the ethnic composition of the Sri Lankan army and the ethnic dimension of the war, it is tempting to ascribe much of this behaviour to straightforward racism. No doubt there is much validity in this assumption. The Sri Lankan military was overwhelmingly Sinhalese and the civilians in the north and the guerrillas overwhelmingly Tamil. The war was also about ethnicity, the Tamils fighting what they saw as Sinhala domination. The inherent tension in the situation was further exacerbated by the language barrier; few if any soldiers spoke Tamil and the army depended heavily on the few Tamil and Muslim soldiers for communication with the civilians[70]. This added to the frustration and anger of the armed forces when the guerrillas melted into the urban or rural landscape after an attack. It was easy for the soldiers to see their enemy as not just a handful of militants but the whole population that was aloof, un-cooperative and conniving with their assailants. This exacerbated the prejudice towards Tamils which many Sinhalese troops and even some of their officers would have already harboured. In July 1983 during the riots in the south many soldiers had

---

[67] Don Mithuna, 'It Will Only be a War of Attrition if They Seek a Military Way Out", *Weekend*, 17. 11. 85, p. 6.
[68] Weaver, 'Civil War Looms With Separatists',.
[69] O'Ballance, *Cyanide War*, p. 38. According to O'Ballance it appeared in the *Times of India* of 7 December, 1984.
[70] Interview with Brigadier Bahar Morseth, 21. 11. 15. There were a handful of Tamil soldiers who were born and raised in Sinhalese dominated areas. The few Muslim soldiers in the army also spoke Tamil.

stood by while mobs went on the rampage and some had even taken part, demonstrating their perception of the Tamil community's responsibility for the militants' attacks. These were men who were removed from the immediate carnage in the North. How much easier would it have been for the men on the spot to assume that connection?

And in the East, where the militants were directly attacking the Sinhala civilians, the temptation to go after the Tamil civilians was even stronger. The Sinhala civilians were not being attacked by Tamil civilians but, in the minds of the mainly Sinhalese security forces, by attacking the Sinhala villagers the militants were inviting reprisals on Tamil villagers. This cruel logic became even more appealing when the massacres served the strategic purpose of protecting Sinhala occupation of the 'border areas.'

However, we must also be careful not to add too much weight to the ethnic factor. Clearly, the ethnic divide between the overwhelmingly Sinhalese security forces and the exclusively Tamil militants played a crucial role in the reprisals against civilians especially in the East where they were calculated assaults on the Tamil civilians and their property. But there were also other strands in the military's atrocities. During the crushing of the second JVP uprising in 1988-89 the military showed that it was no less capable of brutalising the Sinhalese as it was of maltreating the Tamils. In fact, the number of killed and missing during the crushing of that uprising was far greater than the numbers killed in the north. Rohitha Munasinghe's memoirs give a chilling account of the calculated, sadistic violence unleashed by the military on their captives in a southern camp during the JVP uprising.[71] These were not Tamils

---

[71] See Rohitha Munasinghe, *Eliyakanda Wada Kandawura*, Colombo: Godage Brothers, 2000.

who were being tortured and butchered but Sinhalese who shared the culture, language and religion of their tormentors and killers. Counterinsurgency expert Tom Marks who spent some time with the STF in Batticaloa in the mid 1980s also felt that the torture that the STF was accused of was little more than beatings and the application of chilli powder to genital areas and no more than what the police practised routinely in Sri Lanka.[72]

No doubt the Sinhalese soldiers often saw the Tamil civilians and the militants as two sides of the same coin and treated them as such. And, to be sure, the STF was accused of much more than torture in the East and with very good reason. However, it is also likely that lack of experience in dealing with an insurgency and frustration with an elusive enemy also played a big role in making soldiers respond to attacks with nervous spontaneity with fatal consequences for the civilians, especially in the north. The soldiers were faced with a hidden enemy who fled as soon as they had killed and wounded. In such circumstances it was natural for the troops, many of whom were raw recruits, to take their anger and frustration out on the civilians who were present.

In an unguarded moment a government official explained what usually happened in very candid terms:

> "The army goes berserk after a terrorist landmine blows up under their jeep. The mine does not kill everyone. Some army men survive and they stagger out of the vehicle half dazed and then keep firing at everyone. People get killed, even innocent people who are just going

---

[72] Marks, 'Sri Lanka's Special Forces', p.36.

to the market because the army men come out in a state of shock. "[73]

The evidence given by a policeman with regard to an incident on Bar Road in Batticaloa on 16th November 1985 is quite revealing in this regard:

> "At a point on the Bar Road, a land mine exploded killing sub inspector Vitharane and three police officers who were in the lead jeep. The jeep had overturned and a door had been blown off and rested on the roof of a house. Simultaneously there were other blasts similar to grenade explosions. At the same time he had seen men running away from the scrub adjacent to the side of the blast by the roadside. At this stage they all got off their jeeps and started firing in all directions following the standard practice......Firing continued for about five minutes and about 300 rounds were fired."[74]

It is interesting that the police started firing in all directions "following the standard practice." A search in the scrub later revealed that eight civilians had died in the shooting.

Many of the atrocities committed on the civilians following attacks on military patrols bore signs of spontaneous, even nervous reaction, much like the responses described by the policeman and the official above. Usually, the attack on the civilians was launched by the soldiers who survived an attack or their colleagues who arrived at the scene soon after. They burnt with the rage of having seen their colleagues killed and maimed by an enemy who could not be found.

---

[73] Don Mithuna, 'It will only be a War of Attrition'.
[74] 'Monitoring Committee on Batticaloa', *Saturday Review*, 18. 1. 86, p. 5.

As appalling as it was, such behaviour was also predictable, given the lack of experience in the security forces with combat of any kind, let alone an insurgency. And being called upon to fight the worst kind of enemy possible, a terrorist-guerrilla outfit that used unconventional tactics, they were finding themselves in a difficult situation. Without experience or training in handling such situations they were reacting spontaneously, especially in the case of the violence in the North. The predicament of the inexperienced Sri Lankan troops can be appreciated by considering how troops considered to be far more experienced and professional have struggled to deal with similar situations. As a member of the French Foreign Legion in Vietnam recalled, 'rape, beating, burning, torturing of entirely harmless peasants and villagers were of common occurrence in the course of punitive operations by French troops throughout the length and breadth of Indochina.'[75] The behaviour of American soldiers in Vietnam and Russian soldiers in Afghanistan and Chechnya offer further examples of such behaviour by professional soldiers when confronted by an invisible enemy that blended with the population. In more recent times the young US soldiers who smashed their way into Baghdad in 2003 were jittery enough to blast to pieces anything and anyone that looked hostile.[76]

In Sri Lanka itself the experience of the IPKF provides some illuminating examples. On many occasions small units of IPKF soldiers came under fire from the Tigers in

---

[75] Henry Ainley, *In Order to Die*, London: Burke Publishing, 1955, p. 30.

[76] In April 2003, the US forces fired on Hotel Palestine in Baghdad housing foreign journalists. When asked why the officer in charge Lieut. Colonel Decamp explained that one of their commanders had just been shot, "and they don't like it when one of their guys get hit, I can tell you." Peter Wilson, *Long Drive through a short War: reporting on the Iraq War*, South Yarra, Victoria: Hardie Grant Books, Australia 2004, p. 182.

built up areas, suffering casualties without ever seeing their attackers testing the patience of the soldiers to the maximum. General Depinder Singh documents one such incident at an IPKF road block in a village in Jaffna on 14th December 1987. A platoon of the IPKF had arrived in the village at dawn and set up a road block at an intersection. While conducting searches and interrogations the platoon was ambushed losing 2 killed and 3 wounded. The LTTE assailants escaped. What happened after the attack is explained by the officer:

> "As a sort of backlash the men's sense of frustration and anger took hold of them and, in order to avenge the death of their comrades and because some of these were obviously LTTE informers, they wanted to line up and shoot all the civilian suspects whom we had rounded up earlier in the day and who were still in the church. It was a difficult situation. The civilians were defenceless and entirely at our mercy. In the heat of the moment any of the men could have shot one of them and then, except by shooting the culprit myself, it would have become impossible to stop the rest of them from following suit. However, by a combination of persuasion and threats, I managed to calm them down sufficiently to avert the crisis".[77]

The recount shows how easy it was for a situation to get out of control, even in an army which was considered far more 'professional' than the Sri Lankan army at this time. Decisions were made on the spot by the soldiers and officers affected and often, nerves threatened to overcome

---

[77] Depinder Singh, *The IPKF in Jaffna*, Trishul Publications: Delhi, 2002, p. 143.

reason. The calmness of the officer in charge defused the situation on the above occasion but it was not always the case. The first major military action undertaken by the IPKF, the operation to capture Jaffna saw Indian soldiers frequently running amok with dire consequences for the civilians. In one notorious incident, Indian soldiers fresh from an ambush and apparently without any control, burst into the Jaffna hospital shooting indiscriminately, killing many patients and several doctors.[78]

Undoubtedly many Sri Lankan officers would have found themselves in similar situations struggling to control their men. On the ground, some officers reacted angrily to the indiscipline of their men. One exasperated officer told a foreign journalist that indiscipline was so rife that sometimes he had to slap his men and lock them up to prevent them from going on the rampage.[79] Others were less assertive. In July 1983 the Major in charge of the Madagal camp had pleaded with his men to remain in camp but the men simply drove out disregarding his pleas.[80] But it is also likely that just as some officers were trying to stop the rot many others were also unwilling or reluctant to control their men for fear of risking a confrontation with them. The junior officers who were in command of the troops were faced with a dilemma. They were closer to the young men who were going berserk than to the officers above them. If they tried to stand between the men and the civilians they ran the risk of losing the loyalty of their men. On the other hand they could not allow the men to go on the rampage. It is likely that the officers decided to let the men take off their steam on occasion rather than risk losing control over them in general. As an officer in Jaffna fort confided in a local

---

[78] Hoole et al., *Broken Palmyrah*,210-281, Attack on the hospital, pp. 265-72.
[79] Weaver, 'Civil War Looms With Separatists'.
[80] Munasighe, *A Soldier's Version*, p. 8.

journalist, sometimes they had to let the men fire back at their attackers even if that meant civilians could get hurt. "We have troops injured and screaming their heads off", he explained. "At least for the morale of the troops we have to fire back."[81] Besides, many of the junior officers may have been sympathetic to the soldiers having had to endure the same predicament as them unlike their seniors.

It is conceivable that the fact that the enemy was ethnically different encouraged this reaction, making it easier for the troops to brutalise civilians who were ethnically and culturally not their own (which was also the case with the IPKF and the Americans and the French in Vietnam). But as we have seen with the security forces handling of the JVP in the late 1980's it would be too simplistic to ascribe such violence to the ethnic factor alone. Little has been done to study the army's behaviour during the JVP uprising but it is likely that many of the soldiers who engaged in atrocities were responding largely out of frustration and impotence in the face of a ruthless and elusive enemy.

Dealing with the counter-violence by the security forces was also an unprecedented challenge for the state. When the Tamil insurgency escalated, the Security Forces began to study counterinsurgency strategy, mainly with the help of British manuals. But the focus was on military aspects such as conducting ambushes, patrols and raids. The emphasis was on destroying the enemy in combat, not on separating the militants from the population. Even the combat instructions were sometimes offered by personnel with no experience or expertise in the area.[82] Israeli and

---

[81] 'Inside the Jaffna Fort', *Saturday Review*, 4.10.86, p. 11.
[82] Raj Vijayasiri, 'A Critical analysis of the Sri Lankan Government's Counter Insurgency Campaign',, Master's Thesis, Fort Leavenworth, Texas, 1990, pp. 37-40.

KMS expertise does not seem to have made much of an impact on the security forces conduct towards civilians. If anything, it is likely that the mass round ups that caused so much hardship to the population in the north may have been heavily influenced by what Israel was practising in its occupied territories.

Interestingly, the enemy-centric counterinsurgency approach of the Sri Lankan security forces had parallels with the approach of many colonial states. Rhodesia is a classic example. In combating the guerrillas of the Zimbabwe African National Liberation Army and Zimbabwe People's Revolutionary Army, the Rhodesian security forces were focused more on killing guerrillas than denying them popular support. Their training was directed towards this end, emphasising the combat aspects of counterinsurgency[83].

In the absence of a clear and effective counterinsurgency strategy the government struggled to form a coherent response to the violence unleashed by security forces on civilians. On the one hand they came down heavily on the soldiers who were found guilty of atrocities and indiscipline. After their rampage in May 1983 the Rajarata Rifles was disbanded and in July 1983, a number of men who took part in atrocities after the Tinneveli ambush were dismissed.[84] The Rajarata Rifles was amalgamated with another recently raised unit, the Vijayabahu Regiment to form a new Regiment, the Gajaba Regiment.[85] Athulathmudali admitted in June 1985 that more than 300 members of the armed forces have been discharged due to

---

[83] See Marno de Boer, 'Rhodesia's Approach to Counterinsurgency: a Preference for Killing', in *Military Review*, November-December 2011, pp. 35 - 45.
[84] C.A. Chandraprema, *Gota's War: the Crushing of Tamil Tiger Terrorism in Sri Lanka*, Colombo: Ranjan Wijeratne Foundation, 2012, pp. 102-4.
[85] *Sri Lanka Army*, p. 635.

indiscipline.[86] At times the government also showed a willingness to admit publicly the excesses of the security personnel as it did after the army rampage in Mannar in August 1984. Minister M. H. Mohamed who headed a fact finding mission to Mannar reported in detail the violence perpetrated by the security forces. [87]

However, the government also tried to gloss over the horrendous brutalities. Minister Athulathmudali, could be dangerously dismissive about military brutality. While promising to investigate claims of military reprisals, he also spurned allegations of attacks on civilians as "rubbish" and vowed to "resign and go home" if the allegations were proven.[88] Along with such denials came the dangerous practice of describing all non-military casualties of encounters between the security forces and the militants as 'terrorists'. No doubt the government was being careful to avoid an international outcry about civilian deaths, particularly from India. But the practice also reflected a large measure of callousness towards the lives of Tamil civilians. A very early indication of this was the passing of the Emergency Regulation 15A on 3rd July, 1983 which gave the security forces the power to dispose of bodies without holding inquests or revealing the identities of the dead which provided the security forces with the legal cover to carry out reprisals.[89] This was further underlined by the imposition of the naval exclusion zone and a prohibited zone around the northern shores which severely inconvenienced the population. It shows that the government was willing to impose collective punishment on the civilians. It even gives the impression that in a broad sense, the government saw the civilians and the militants being the same and therefore deserving the harsh treatment. Such

---

[86] Lalith Athulathmudali's interview with Qadri Ismail *The Island*, 16. 6. 85, p. 9 .

[87] 'Army 'Guilty' in Sri Lanka', *Herald*, 16. 8. 84.

[88] William Clairborne 'Tamils Hide in Fear as Troops Take Revenge', from the Washington Post, published in *The Age*, 15.6.84.

[89] Hoole et al, *Broken Palmyrah*, p. 58.

attitudes could only encourage the soldiers in their quest for revenge from the civilians.

One can, to an extent, appreciate the problem faced by the government. It had never faced a situation of having to deal with security forces that were both under intense pressure from an enemy and seemingly out of control in relation to the civilians. But the government's response, in particular the harsh measures of control imposed on the civilians in the north showed that it also had little interest in dealing with the problem as a complex political issue. The security measures were very much a knee-jerk reaction to the violence on the ground rather than the underlying causes of it, aimed at making the security forces safe and accountable without attempting to understand the nature of the problem they had to deal with.

But it only made the role of the security forces harder. The civilians whose homes were burnt and whose loved ones were massacred and beaten had little sympathy for the army. 'We are despised here." One officer confided in a foreign journalist. "We just can't cope with the situation". [90]The former chairman of the Velvetithurai urban council, an area which had received much punishment from the security forces, spoke for many Tamils in the north. "The Sinhalese army, far from ensuring our security is an object of terror for us" he said. 'We now go to the Liberation Tigers for help whenever threatened."[91] Tamils, driven into the arms of the militants by the July riots, were being further alienated by the government that failed - or was reluctant - to reign in its armed forces.

In facing the Tamil insurgency then, the Sri Lankan military showed all the signs of an inexperienced, poorly

---

[90] Weaver, 'Civil War Looms With Separatists'.
[91] S.H.Venkatramani, 'Battle Lines', *India Today*, 15. 12. 84.

equipped and ethnically homogenous army trying to come to terms with a deadly insurgency waged by a committed guerrilla force belonging to a different ethnic group. The government's failure to appreciate the need for an effective counter-insurgency strategy complicated matters. As a consequence the military's its response to the challenge of Tamil insurgency was generally nervous, unprofessional and more damaging to the civilians than the militants.

CHAPTER 4

The Siege

On June 18 1985, the Sri Lankan government and the Tamil militants agreed to a ceasefire, the first major cessation of hostilities between the two sides. The ceasefire was largely the result of lobbying by the Indian government to bring the two warring sides to some agreement. India did not want the conflict in Sri Lanka to spiral out of control and wished to see it settled in a way that safeguarded its strategic interests. Following the ceasefire, talks were to be held between the militants and the Sri Lankan government in Thimpu, the capital of Bhutan. But before this, India wanted to bring the different militant groups together so that they may form a united front vis-a-vis the Sri Lankan government. After some wrangling, the groups agreed to form the Ealam National Liberation Front (ENLF), and the talks began on 8 July.

4.1 Talking Peace, Making War

The Thimpu talks collapsed in August after two rounds. The ceasefire held on until the following year. It was tenuous at best with both sides accusing each other of numerous violations. In the Jaffna peninsula it remained more or less intact but in the east the killings continued. Then, on 13th January 1986 the Tigers launched a barrage of mortar fire on the Jaffna fort and the ceasefire was finally, officially over. From then on the two sides slid almost unstoppably towards a major showdown.

The collapse of the Thimpu talks demonstrated the bleak prospects for a peaceful settlement to the conflict. The war, at this stage, was being dominated on both sides by a desire to try the military option. Although the conflict was taking a steady toll in lives and affecting the lives of tens of thousands of others directly and indirectly, a peaceful settlement to the conflict had never been a serious pursuit for either the government of Sri Lanka or the Tamil militants. War, both sides believed, was a surer option for achieving their goals than peace despite its attendant destruction. The collapse of the talks and the break-up of the ceasefire was very much a natural outcome of this attitude.

The government's position on the conflict was heavily influenced by the outlook of the majority Sinhalese community on which it depended for political power. The overwhelming view in the Sinhalese south was that the militancy in the north was terrorism and not the manifestation of genuine Tamil grievances. The TULF, the mainstream Tamil party was seen as an ally of the militants at best, their puppet at worst. In this context, any attempt to devolve power was a step, no matter how minute, in the direction of Ealam which the 'terrorists' and the TULF were standing for. Whether the Sinhalese majority would have countenanced any attempt to devolve power to the Tamils even without the existence of a militancy associated with the demand for a separate state is a moot point, considering their vehement opposition to devolution in 1958 and 1965; however, the militancy only served to harden this attitude even further.

It was mainly Indian pressure that kept the Sri Lankan government on the path to seeking a political settlement. Following the riots in July '83 and the escalation of violence in late 1984 the spectre of an Indian invasion was raised by the local media. In January 1985 in the wake of the militants' 'offensive,' the ruling UNP went so far as to accuse India of using the Tamil militants as 'proxy troops' in an invasion of Sri Lanka.[1] India was certainly not interested in an invasion directly or by proxy, but as we have seen previously, India's interest in Sri Lanka was serious enough to provide the militants with arms and training. But while maintaining pressure on the Sri Lankan government with such covert operations India also exerted pressure for a political solution. In late 1983 Gopalaswamy Parthasarathy was sent to Sri Lanka to broker a peace deal resulting in the All Party conference (APC). The APC met throughout 1984 but produced little. Then in 1985 with the militants on top, India persuaded the main militant groups to form the ENLF and participate in talks with the Sri Lankan government in Thimpu.

The consequence of this was that the Sri Lankan state and the Sinhalese people saw India as part of the problem and a major part at that. The giant northern neighbour was seen as trying to endanger Sri Lanka's sovereignty in collusion with the 'terrorists,' preventing the armed forces from defeating the militants. Such perceptions did much to define the government's approach to dealing with the insurgency as illustrated by the harsh measures taken against the civilian population in the north. Force, rather than winning over the civilians was seen as the key to dealing with a terrorist problem which had its main source of support across the Palk Straits. Again, the parallels with Rhodesia are striking. The White government in Rhodesia saw little point in dealing with Black discontent because

---

[1] Michael Richardson, 'Sri Lanka Accuses India of 'Invasion', *The Age*, 31. 1. 85.

they believed the insurgency in Rhodesia was fomented by external powers such as China and the Soviet Union.[2]

For their part the militants too saw little merit in peace talks. They had the government forces with their backs to the wall and they wanted to press harder when they had the advantage. The peace talks undermined this drive and they felt cheated by India. Thus, while the Sri Lankan delegation went to Thimpu with the determination to find a solution within the existing constitutional framework which was wishful thinking, the militants went there determined to expose the government's duplicity rather than to seriously work out a solution, as they were convinced that the government was not interested in one. It is also arguable that, as J. N. Dixit has claimed, that the militants believed that by demonstrating the GOSL's intransigence they could persuade India to offer more substantial help towards achieving their goals.[3]

With neither side ready for serious discussions, the talks collapsed. The failure of Thimpu talks led to a renewal of fighting in the north. But now the ground situation had changed. The Tigers had utilised the ceasefire that came into effect before the talks to strengthen their bunkers around the army camps and ring these bases with minefields. They had placed sentries, usually young boys, all around the camps to watch the soldiers. Whenever a patrol tried to move out the sentries alerted the main forces of the militants who rushed to the place and attacked the enemy patrol with grenades and gunfire.[4] Fortified bunkers and outposts sprang up in close proximity to military

---

[2] Marno de Boer, 'Rhodesia's Approach to Counterinsurgency: a Preference for Killing', in *Military Review*, November-December 2011, p. 36.

[3] J. N. Dixit, *Assignment Colombo*, Colombo: Vijitha Yapa, 1998, p. 43.

[4] Malini Parthasarathy, 'A Military and Political Misadventure', *Frontline*, May 31 – June 13 1986, p. 19.

encampments, almost taunting the soldiers inside. From behind walls of empty houses or piles of sandbags, young rebels watched their enemy, clutching their rifles and keeping a few grenades on stand-by. All roads leading to the Army camps were plugged with road blocks, built of burnt out cars, old tyres, and sandbags. In some places even walls had been constructed across the roads.[5] The blockade had begun even before the ceasefire but now the circle seemed to have grown closer, the noose tighter. A cocky young guerrilla leader boasted to a foreign correspondent that the army was now fully under control. Whenever they tried to move out, the leader assured, the guerrillas were able to "chase them back."[6]

As 1986 progressed the situation changed even further. When the Thimpu talks collapsed the barricades and bunkers were had been manned by cadres of all the major armed militant organisations, but by the middle of 1986 they were exclusively from the LTTE and EROS. The reason for this was a bloody power struggle that brought the LTTE to the helm of affairs in the north. The Tigers' leader Prabhakaran had never been very happy with the other militant groups, particularly TELO which he considered an Indian stooge. With considerable support from India, TELO had built up its military machinery while the LTTE had boosted its own power largely independent of India's aid. It was soon clear that Jaffna was not big enough for the two rival groups. Using a street clash between its cadres and TELO men, the LTTE launched a lightning strike on the TELO on 29th April 1986. Tiger cadres pounced on all the TELO camps and hideout in the Jaffna peninsula ruthlessly gunning down their rivals. Some were killed in their sleep,

---

[5] S. Venkatnarayan, 'A Visit to Jaffna', *The Island*, 15. 2. 87, p .6.
[6] Michael Hamlyn, 'The Boys' Keep Army Behind Barricades in Fight for Tamil State, *The Australian,* 22. 3. 86.

some while having meals. The TELO leader Sri Sabaratnam escaped and went into hiding but the Tigers tracked him down and killed him on 7th May. Not stopping at murdering their rivals the LTTE went on to make a telling example of its brutality to anyone daring to question its primacy. Some of the victims were piled up at road junctions and burnt in public. At least one was thrust into a car which was exploded. Sabarathnam's bullet riddled body was displayed at the Jaffna bus station to demonstrate to the population the end of the TELO[7]. 'Like mad dogs they killed our men', a survivor recalled with horror.[8]

Within a week it was all over. TELO had ceased to exist in all but name. Other small groups of militants were quickly absorbed into the LTTE.[9] In October, the LTTE ordered the PLOT to leave Jaffna. Impotent before the almighty Tigers, the group complied but not before their Jaffna commander was brutally murdered by the Tigers.[10] Of the other groups, the EPRLF retreated to Batticaloa where lay their main base while EROS alone established a working relationship with the LTTE. Hereafter it was to be the LTTE that called the shots in the Tamil struggle.

And to make the point clear, the LTTE also brought the war to Colombo and the environs for the first time. On 3rd May a bomb exploded aboard Air Lanka flight UL512, breaking the air craft in two and killing 21 people including 13

---

[7] M. R. Narayan Swamy, *Inside an Elusive Mind: Prabhakaran*, Delhi: Konark Publishers, 2003, pp. 135-6, Rajan Hoole et al., *Broken Palmyra: The Tamil Crisis in Sri Lanka, an Inside Account*, Ratmalana, Sri Lanka: Sri Lanka Studies Institute, 1992, pp. 81-2.
[8] Jehan Hanif, 'Massacre in Jaffna: TELO Man Tells All', *The Island*, 14. 9. 86, p. 13.
[9] Edgar O'Ballance, *Cyanide War: the Tamil Insurrection in Sri Lanka 1973-88*, London: Brassey's, 1989, pp. 61-2.
[10] Narayan Swamy, *Inside an Elusive Mind*, pp. 141-2.

foreigners. The explosives had been hidden in a container with meat, vegetables and fruits destined for Male. The explosion was to occur in mid air but a 45 minute delay resulted in the bomb ripping through the aircraft while passengers were still boarding.[11] On 7th May 1986 a parcel bomb ripped through the Central Telegraph Office in the Fort area in Colombo killing 11 and injuring more than a hundred. The colonial era building suffered severe damage, its wooden upper floor collapsing.[12]

By the latter months of the year the Tigers were well in control of the peninsula. By the end on 1986 they had begun to levy their own taxes. They had also set up a court system to try criminals and anti social elements.[13] Prabhakaran was still in South India at this time but under the leadership of Sathasivam Krishnakumar or Kittu who was in charge in Jaffna, the Tigers meted out rough justice to those who disobeyed them. Many who were found guilty were executed and displayed tied to lamp posts.

With Jaffna under their control, the Tigers intensified their training and arms production with a view to escalating the pressure on the army bases in the north. They had established five main camps around Jaffna where new recruits were being trained. An increasing number of them were boys, some in their early teens. But unlike in 1984/85 the Tigers were now struggling to meet the growing

[11] T. D. S. A. Dissanayake, *War or Peace in Sri Lanka*, Colombo: Popular Prakashan, 1995, vol. 2, p. 71.

[12] William de Alwis and Srimal Abeywardene, 'Eleven Killed by Bomb in CTO', Ceylon Daily News, 6. 8. 09, http://www.dailynews.lk/2009/06/08/fea10.asp

[13] Michael Hamlyn, 'Tiger Guerrillas Step into Rulers Role: Jaffna Tamils Prepare for Post-settlement Role in Sri Lanka', *The Times*, 18. 9. 86. The rural court in Manipay consisted of the areas elders and distinguished citizens. It met three times a week. S. Venkatnarayan, 'A Visit to Jaffna'.

demands of their military commitments, having to rule the peninsula and besiege the military camps all on their own.[14] The five camps were reportedly producing around 80 fighters every three months. By now every graduate of Tiger training was wearing a 'cyanide necklace', which simply consisted of a phial of cyanide hanging from a string around the neck. [15] The rebel cadre was expected to bite into the phial and commit suicide if faced with capture. The cyanide necklace would soon become an accoutrement symbolic of the LTTE's commitment to its cause.

In an effort to give their forces an aura of professionalism the Tigers also began to confer regular military ranks on their cadres. Some of the regular cadres wore military style uniforms but the majority of the Tiger fighters still wore civilian clothes. Few, if any, wore shoes, the fighters preferring rubber slippers or sometimes going about just in their bare feet. From late 1985 the Tigers also began inducting women into their fighting ranks. They were formed into separate female units but initially placed under male command, the Mannar LTTE commander "Lt. Colonel" Victor taking charge of them. Under Victor's leadership the female cadres had their first taste of battle in Adampan in October 1986 when they fought along with male Tiger fighters to stop a military encirclement of the area.[16]

By now the Tigers' weapons 'factories' were in full swing manufacturing weapons and ammunition. They manufactured mortars and shells and even the ammunition for RPGs. The mortars were 50, 81, 90 and 155mm and had a cast iron or aluminium shell and black powder as the

---

[14] Hoole et al., *Broken Palmyrah*, p. 78.
[15] O'Ballance, *Cyanide War*, p. 66.
[16] Adele Balasingham, *Women Fighters of Liberation Tigers*, Jaffna: Thasan Printers, 1993, pp. 17-35.

propellant.[17] In October 1986 Kittu boasted that the Tigers were producing around 25 mortars and 100 grenades daily[18]. "We prefer our own mortars to the ones we have bought" he declared proudly.[19] The mortars, some of which were as big as 155mm, had aluminium cases which according to the army, caused shrapnel to explode in large pieces causing horrible wounds to the soldiers.[20] A visiting Indian journalist however, was not impressed by the Tigers 'ordnance factory' calling it 'nothing more than a junk shop'.[21] Such contempt for the home-grown technology of the rebels was justified to an extent by the performance of the products of these 'factories'. The mortars fired at the Jaffna Fort were frequently inaccurate, sometimes only about a third of those fired hitting the Fort.[22] But it was sufficient to keep the soldiers cooped up and ducking for cover.

But the Tigers were also making use of more sophisticated weapons. Now there were more rocket launchers around, usually the soviet RPG-7. Tiger cadres were now armed with a variety of rifles, the commonest being the AK-47, T-56 and M-16 rifles. By mid-1986 they had also purchased a small number of Browning .50 calibre machine guns. These were mounted on pick- up trucks for easy mobility. Their firepower was supplemented by several Browning .30

[17] Rohan Gunaratna, *War and Peace in Sri Lanka*, Colombo: Institute of Fundamental Studies, 1987, p. 47.
[18] Dexter Cruez, 'Within the Jaws of the Tiger', *Weekend*, 26. 10 . 86, p. 8
[19] Jon Swain, 'Face to Face with the Guerrilla commander: Cyanide Martyrs bar way to peace', *Sunday Times*, 10. 8.86, reproduced in *Lanka Guardian*, 1.9.86, pp11-12.
[20] 'Inside Jaffna Fort: Battle of Nerves', *Ceylon Daily News*, 11.9.86, reproduced in *Saturday Review*, 20.9.86, p. 6.
[21] Venkatnarayan, 'A Visit to Jaffna.'
[22] 'Tigers Hold Troops as Captive Force', Reuters report, *Weekend*, 21. 6. 87, p. 11.

machine guns.[23] There was also at least one 40mm grenade launcher in use. These were supplemented by improvised grenade launchers which consisted of ordinary hand grenades fired from sawn-off shot guns.[24]

4.2 Under Siege

For the security forces, the tightening of the siege and the emergence of the Tigers as the predominant guerrilla group posed a serious challenge. The blockade meant that troops were now more or less stuck inside their camps, dependant on supply by air and sea while the rise of the LTTE placed the camps at the mercy of the Tamil militant group that was most committed to eliminating them.

In order to understand the predicament of the troops in Jaffna at this time we need to look at the strategic geography of the Jaffna peninsula. As explained earlier, the Jaffna peninsula is connected to the mainland only by a narrow isthmus at its south eastern end and the main artery that links the peninsula with the rest of the island crosses the Jaffna lagoon at Elephant Pass, to the west of this isthmus. This road, the A9, ran all the way to Kandy through the Vanni and was also the main supply route to the civilians and the troops in Jaffna. By early 1986 there was minimal military presence along this route or in the jungle clad expanse of the Vanni. The small Sri Lankan army was already too stretched in the northern peninsula and parts of the east to occupy the ground in the Vanni in

---

[23] Gunaratna, *War and Peace in Sri Lanka*, pp.46-7, Thomas A. Marks, 'Counterinsurgency in Sri Lanka: Asia's Dirty Little War', *Soldier of Fortune*, Feb. 1987, p. 42.
[24] Interview with Brigadier Bahar Morseth, 21. 11. 15. The Tigers removed the safety clip of the grenade and held the lever in place by means of a rubber band that went around the grenade. The base of the grenade was then fitted to a stick which was fired from a sawn-off shot gun.

strength. There were only two outposts between Vavuniya and Mankulam and only the small post guarding the TV and radio transmitter at Kokavil between Mankulam and Kilinochchi. On the road from Mankulam to Mullaitivu on the east coast there was only a small camp manned by an engineering squadron at Oddusuddan. There was virtually no military presence between Vavuniya and Mannar. This allowed a vast area in the Northern Province to go under the control of the rebels without a fight. There was very little the army could do, tied up as they were in the north. But as in the peninsula, here too the army was capable of holding out in their camps. This was shown when on 4th June 1986 the Tigers launched a sustained attack in force on the Kilinochchi camp. About 400 Tigers are supposed to have taken part in the attack. For five days the militants laid siege to the camp, preventing the soldiers from coming out. But the army held out, beating off the attack only with the help of air support. Reinforcements by air finally forced the Tigers to abandon their efforts on the 9th.[25]

But outside their camps, on the long, jungle fringed roads, the army was highly vulnerable to guerrilla attacks. As the army had no tracked vehicles movement had to be confined to the roads which made them vulnerable to ambushes. The troops usually travelled these roads in large convoys, sometimes of dozens of vehicles with armoured escort and air cover. Still, the guerrillas could launch well-coordinated attacks that took a heavy toll on them. One such attack took place at Kokavil on the Vavuniya - Kilinochchi road on April 2nd 1986 when the Tigers sprang a well-organised ambush on a military convoy. According to some reports they set off two mines at either end of the convoy to trap it

---

[25] O'Ballance, *Cyanide War*, p.64, *Sri Lanka Situation Report*, Tamil Information and Research Unit, Madras, no. 15, (15. 7. 86), pp. 5-6 and no. 16 (1. 8. 86), p. 3.

and then opened fire on the vehicles which included trucks and armoured cars. A series of grenade explosions was also set off along the length of the convoy on a 600 meter stretch of the road before they attacked the stalled vehicles with rockets and small arms. Seven soldiers were killed and 15 wounded while the LTTE admitted to losing 4 fighters.[26] And wherever possible, the Tigers also attempted to enforce more direct control over the A9. Between Kilinochchi and Elephant Pass they set up check points and road blocks making it impossible for the Security Forces to use that stretch of the highway. [27] As a consequence, even though the A9 was not completely cut off, using it as a main supply route was becoming increasingly difficult.

The restricted mobility of the troops along the main arteries of the Vanni meant that large areas of the region were virtually ceded to the rebels. The Tigers used this to good advantage. The coast around Mannar in particular was crucial for the Tigers' connection with India. They inserted men and material from Tamil Nadu into this area and then taking advantage o the minimal Army presence in the interior, moved them unimpeded through the jungles to their northern bases. On the east coast similar insertions took place north of Trincomalee. The vast forested hinterland of both coasts was connected to the coast by a string of jungle bases that functioned with little fear of intervention from the security forces.[28]

With the road link tenuous, the garrisons in Jaffna were now almost totally dependent on air and naval power for supplies. At this time there were two main military

---

[26] Humphrey Hawkesley, 'Tamil Guerrillas kill 7 in Troop Convoy Ambush', *The Age*, 3. 4. 86, *Sri Lanka Situation Report*, Tamil Information Research Unit, Madras, no. 9 (15. 4. 86), p. 2, and no. 11 (15. 5. 86), p. 5.
[27] Iqbal Athas, 'Operation Giant Step', *Weekend*, 22. 2. 87, p. 7.
[28] Ibid.

encampments in Jaffna, the base at Palali with its airstrip and that in the Jaffna fort in the south western corner of the peninsula. There were several other smaller detachments at Point Pedro, Velvetithurai, Elephant Pass, Pooneryn and Navatkuli. All these camps were located in places that could be easily accessed by sea or air. With the A9 all but cut off, they were virtually marooned from the rest of the island.

And the post-Thimpu blockade meant that troops were now not only isolated from the rest of the island but also confined within the perimeters of their camps in the peninsula. Even patrolling beyond the perimeters of the camps now was to invite serious confrontation. Often, the patrols had to return as soon as they had left. Some returned without loss but others received terrible mauling from the Tigers.[29] Occasionally a patrol would complete its mission of covering a certain area but at a painstakingly slow pace as they had to comb the route for landmines and be on the lookout for snipers and other ambushes. One journalist reported a convoy of several Saracen APCs and Saladin armoured cars taking an hour and ten minutes to cover the six miles between Palali and Kankesanthurai[30].

Soldiers were not completely secure inside their bases either. Mortars and RPGs had come to play a significant role in the Tiger arsenal by now and they regularly pounded the camps with mortar shells and rockets while firing small arms at any sentries daring to show themselves. The objective was to keep the troops pinned down and edgy. Occasionally a lucky strike could kill and maim as an unfortunate Lieutenant in the Jaffna Fort found on his 24th

---

[29] Tim Smith, *Reluctant Mercenary, the Reflections of an Ex-Army Helicopter Pilot in the Anti-Terrorist War in Sri Lanka*, The Book Guild Ltd., Sussex 2002, p. 71.

[30] Iqbal Athas, 'My Longest Day', *Weekend*, May 11, 1986, p. 6.

birthday on July 28th 1986. As he turned up at the mess to celebrate his birthday with his colleagues a mortar crashed through the roof, killing him and a rifleman instantly. The fort was a favourite target of the militants, a few mortars finding their way within its walls almost every night. An officer quipped that mortar fire was so frequent that they fell like rain.[31] This was no doubt an exaggeration but when the Indian army arrived in July 1987 they found plenty of signs of the battering received by the fort with many of the buildings inside virtually turned into rubble.[32]

For the troops in the northern camps the siege was a terrible strain. The soldiers were initially posted to Jaffna for a year but this was later cut down to six months due to the stress they suffered. The only link with the outside world was the Air Force flights which flew twice daily bringing reinforcements, fresh vegetables and newspapers.[33] Usually the night was spent inside bunkers to escape the intermittent mortar fire, soldiers swapping stories about girlfriends or watching videos of war movies with little to look forward to the following day other than more of the same. It was a miserable existence that made a tour of duty in Jaffna seems like a prison sentence[34].

Lieutenant Colonel Modestus Fernando remembers the experience of being posted to Jaffna during these stressful times. Upon completing his officer cadet training, Fernando and his colleagues were posted to Jaffna Fort on their first tour of duty. Modestus arrived in Jaffna still wearing his training insignia on his sleeves and on the first night was taken around the fort by a senior officer. They were shown

---

[31] Iqbal Athas, 'The Fear of Living Dangerously', *Weekend*, 3. 8. 1986, p. 23.
[32] Depinder Singh, *The IPKF in Jaffna*, Delhi: Trishul Publications, 2002, p. 46.
[33] Venkatnarayan, 'A Visit to Jaffna.'
[34] 'Inside Jaffna Fort: Battle of Nerves'.

the layout of the camp, the strong points as well as the places in the surrounding area from which the enemy sniped at them. It was a surreal experience for the young officer, walking around defences that had turned the 17th century Dutch fort into a strange landscape. From the following night the action started. Mortars began to fall intermittently, introducing the young officers to the reality of combat. The Tigers had christened their heavy mortars 'Baba' on account of their size which was that of a toddler or 'baba.' It was an unnerving sight watching them approach, swaying in the wind without a clear indication of where they might land.[35]

In the East the situation was somewhat different. The presence of Sinhalese civilians, especially in Trincomalee, led to the army establishing a strong presence around them by setting up numerous camps. In the whole of the Eastern Province, by early 1986, the security forces had established as many as 49 camps in strategic locations of the road network to create a mutually supporting web of strong points.[36] The objective was to keep the area well covered and to make it difficult for guerrillas to carry out any attacks within the mesh of camps as it would have trapped them. This heavy military presence, and the fact that the rebel forces were thinner on the ground in the East than in their northern stronghold, enabled the security forces to move about more freely, at least during the daytime. As a result they were able to carry out frequent raids on suspected militant camps and strongholds. In Batticaloa the STF was in charge of operations, trying to keep the enemy on their toes. Operating mainly in sections of twelve men roaming around in land rovers and APCs, they carried out

---

[35] Interview with Lieutenant Colonel Modestus Fernando, 7. 8. 2016.
[36] O'Ballance, *Cyanide War*, p. 58.

routine checks, round ups and ambushes.[37] The urban areas were generally deemed secure. But people remained cautious; the Trincomalee town was reportedly deserted by noon.[38]

However, whilst the siege in the north was a serious setback for the security forces, there were also encouraging signs for them. During the ceasefire, they had also received a boost to in terms of training and new armaments. The batch of cadets that passed out with Modestus Fernando had returned from training in Pakistan. They were the first such batch to have received such training as part of a program offered by Pakistan to train Sri Lankan officer cadets from 1985. The training was gruelling often with 2 AM starts and the cadets were trained in a variety of fighting skills and techniques suited to jungle and guerrilla warfare.[39] New weapons were also coming in. President Jayawardena had made trips to China and the United States in 1984 looking for support to strengthen the armed forces. As explained above these trips yielded little. But undaunted, Athulathmudali sought help from sources that were willing to supply what Sri Lanka wanted – and could afford – discretely. These efforts began to bear fruit soon. In the second half of 1985 arms and equipment began to arrive in large quantities. These included 25 pounder artillery, light aircraft, helicopters, recoilless rifles, RPGs, .50 machine guns and large quantities of ammunition.[40] By

---

[37] Thomas A. Marks, Sri Lanka's Special Forces, *Soldier of Fortune*, July 1988, pp. 32-9 (passim)

[38] Iqbal Athas, 'How Forces React to the Terrorists' Grand Design', *Weekend*, 28.9.86, p.10 and 18.

[39] Interview with Modestus Fernando.

[40] The 25 pounders, recoilless rifles and RPGs came from Pakistan while at least 15,000 small arms were reportedly bought from Singapore, Pakistan and China. V.G. Kulkarni, 'The Military Modernises to Meet Rebel Threat', *Far Eastern Economic Review*, 12. 6. 86, pp. 29-31. *Sri Lanka Army*, p. 393.

mid 1985 two plane loads of mortars and RPGs had also arrived from Pakistan.[41] The Army also received an unspecified number of Saladin armoured cars from Britain and new Buffel APCs from South Africa.[42] The latter vehicles with their mine deflecting chassis boosted the confidence of the soldiers to some extent. Later, the electrical and mechanical engineers of the Sri Lanka army also started work on locally produced armoured vehicles based on a Japanese truck chassis and modelled on the Buffel. Called Unicorns, these APCs were an additional boon to the embattled soldiers.[43] The Air Force was also expanding its fleet, acquiring more Bell 212s and Bell 412 by late 1985. Fitted with machineguns and rocket pods these boosted the small fleet of armed helicopters. More significantly, several Siai Marchetti SF260 jet trainers had also arrived in late 1985. These light aircraft, used in a counter insurgent warfare by several other countries, were capable of carrying 250-lb bombs, 70mm rocket pods and machine gun pods and could be used in a ground attack capacity.[44] With these acquisitions, along with the beefing up of the naval arm with Dvoras the Security Forces felt it now had the muscle to pack a potent punch.

The army too was becoming bigger, better armed and organised. By 1986 the numbers had risen to nearly 30,000, a result of the continuing intensive recruiting campaign following

---

[41] Robert Karniol, 'Rocket Boost for Sri Lanka', *Jane's Defence Weekly*, 28. 6. 2000. The TELO bulldozers that led the attack on Kokavil in May 1985 were stopped by RPGs, the first to be fired by the Army in Sri Lanka. *Sri Lanka Army*, p. 428.

[42] O'Ballance *Cyanide War*, p. 55 (Saladins) and p. 58 (Buffels). The Saladins were received in February 1985.

[43] Brian Blodgett, *Sri Lanka's Military: the Search for a Mission*, Aventine Press: San Diego, California, 2004p. 92.

[44] 'Tigers and Lions in Paradise: the Enduring Agony of the Civil War in Sri Lanka', http://worldatwar.net/chandelle/v3/v3n3/articles/srilanka.html

the escalation of the fighting in late 1984. True, there were still many drawbacks. There was still no one standard rifle for the infantrymen, many of the troops being armed with T-56 assault rifles but carrying the Belgian FN FAL rifle and even a few Soviet-made AK47, a part of a stock received in 1971. The troops still lacked transport. More and more Unicorn and Buffel APCs were entering service along with the existing fleet of Saracens, but this was still not sufficient. Civilian vehicles were still being commandeered for troop movements. 'Armour' still consisted of a fleet of Saladin Armoured cars and Ferret scout cars. Infantry units were allocated armour from this pool when operations were carried out. Still, there were some notable improvements. The army's firepower had been markedly enhanced by the addition of the 25 pounders to the aging collection of 76mm and 85mm field guns. The infantry was also getting more streamlined with increased firepower. Each battalion was split into several companies each comprising four platoons of 30 men. The platoons were further broken in to 10-man sections. Nine men in each section were armed with rifles while the remaining soldier carried a light machine gun. One of the riflemen also carried a rocket launcher, usually a Chinese copy of the RPG-7. In some units this was replaced by the West German Heckler and Koch Grenatenpistole 40mm grenade launcher or the South African made Amscor six shot grenade launcher. Sometimes a 60mm mortar was included for extra support. Thus armed the Sri Lankan army now was a much more formidable force than it was in the early 1980s.[45] The STF was also similarly armed and operated mainly in sections of twelve men.[46]

---

[45] Thomas A. Marks, 'Counter-insurgency in Sri Lanka: Asia's Dirty Little War', *Soldier of Fortune*, February 1987, p. 42, Blodgett, *Sri Lanka's Military*, pp. 92-3.
[46] Marks, 'Sri Lanka's Special Forces', p. 34.

The army was also exploring ways to engage the enemy more aggressively. In early 1986 they began working on a "tracker" team. Formed under the command of Major Gamini Hettiarachchi, this was simply a long range reconnaissance team that conducted small group operations deep inside enemy territory. Originally comprising 2 officers and 38 men they were based in Vavuniya and were equipped with specialised equipment like sniper rifles and trained to live away from base for days. They were deployed in the East where the ground situation was more conducive to such operations. They impressed their commanders very quickly. In one operation, a team of four led by Captain Suresh Hashim landed by boat south of the Trincomalee Bay and after lying in ambush for several days, killed the Trincomalee rebel leader and escaped under heavy fire into a waiting helicopter, carrying the body of the slain rebel with them![47] The project was still at an experimental stage with numbers being small but it showed that the military was beginning to think in new ways to undermine the enemy. The commando squadron, of which the tracker team was a part, was itself elevated to the level of a separate unit in March 1986. [48]

In the East, there were also signs of a more nuanced approach to counterinsurgency. In Muttur the forces under the leadership of Major Mohan ,, one of the very few Tamils in the army, carried out aggressive patrolling and positive engagement with the community, clearing parts of the area of guerrillas. Rockwood's A Company dominated the area assigned to them with patrols and ambushes and won the trust and support of the community by helping them rebuild their lives shattered by the war. Rockwood, a fluent speaker of both Sinhalese and Tamil,

---

[47] Cyril Ranatunga, *Adventurous Journey*, Colombo: Vijitha Yapa, 2009, p. 118.
[48] *Sri Lanka Army*, p. 391, Smith, *Reluctant Mercenary*, p. 217.

also encouraged his men to learn Tamil in order to deal with the civilians. As a result they succeeded in killing a large number of guerrillas and clearing a substantial area of rebel influence.[49]

Such experiments were still largely in their infancy and reflected more the genius and ability of individual officers than a coherent strategy. Overall, the focus was still very much on the combat aspect of counterinsurgency. And the expansion of the security forces and the arrival of new weapons only boosted the confidence of the armed forces, increasing the urgency to use them in battle. The improving capability of the armed forces was not lost on the president who told the BBC in October 1985 that militants could be defeated in a year.[50] In January the following year he promised to offer a political solution but only after the military issues have been dealt with.[51] The focus seemed to be clearly on the military front.

Still, the initial response of the security forces to the developing siege in the north was defensive. With the breakdown of the Ceasefire the anxious army declared a 1000 meter security zone around Jaffna fort and other military encampments in the north to deter the militants from using the built up areas around the camps to stage mortar attacks. It demonstrated again that the security forces placed little importance in winning the goodwill of the people. The security zone forced the evacuation of many businesses and residences in the vicinity of the camps, sometimes the bulk of the populations of towns like Velvetithurai and Point Pedro had to be relocated to 'safer' areas to accommodate the army's security concerns. The army, however, was firm. Apologising for the inconvenience to the residents, the commander of the Jaffna Fort Brigadier Rupesinghe said

---

[49] Marks, 'Counter-insurgency in Sri Lanka', pp. 38-9.
[50] 'We Will Defeat the Terrorists in One Year: JR Tells BBC', *The Island*, 27. 10. 85, pp. 1 and 2.
[51] *Military Problem'*, *Saturday Review*, 1. 2. 86, p. 1.

pointedly that no sooner the requirement cease the security zone too will cease to exist."[52] But the security zone was far from a success. It only generated more resentment towards the army without decreasing the attacks on the garrisons appreciably or making patrols safer. Moreover, by pushing the civilians further from the camps in to rebel dominated areas the army made it easier for the enemy to tighten their control over the population. Clearly, more foresight was required in devising strategies for relieving the military pressure.

But the main focus was now shifting to taking the offensive. With peace talks collapsed and new weapons in, the urge to take the battle to the enemy was strong. But again, civilians posed a problem. The military camps were still located in areas close to civilian settlements and any foray from these camps ran the risk of causing heavy civilian casualties. The camps of Jaffna and Palali in particular were at either end of a thick urban corridor inhabited by tens of thousands of people. Venturing into such an area in a major operation was also fraught with serious risks for the troops as the closely built up nature of the area could be exploited by the guerrillas.

To deal with the challenge of urban guerrilla warfare Sri Lankan security forces turned to their new-found ally Israel for help. Under the tutelage of Israeli officers and other ranks non-commissioned officers and junior leaders of the Sri Lanka Army underwent training in close order battle in built up areas at camps in Maduru Oya and Minneriya. This was the first time such training had been imparted to the Sri Lankan army. At Maduru Oya a mock town was constructed with facades of buildings to give the feel of an urban environment while live ammunition was used during

---

[52] 'Jaffna Under Siege', *Saturday Review*, 25.1.86, p. 1.

the exercises. Troops were trained for one month and then deployed in the north in anticipation of action.[53]

And action was becoming imminent now. The internecine war between the TELO and the LTTE provided the government with too good an opportunity to squander. The enemy, it seemed was tearing itself apart. Then came the bombs in Ratmalana and the Central Telegraph Office. The attacks incensed public opinion in the south bringing increasing pressure on the government. The iron, it seemed, was hot enough to strike.

In the event however, the long awaited push became anti-climactic. After days of military build up in May 1986 the security forces launched two substantial operations, dubbed Short Shrift I and II to ease the pressure on some of the military bases in the Jaffna Peninsula. The army, supported by air and naval cover, made a multi-pronged advance into rebel held territory on 18th May. The troops advanced from Velvetithurai, Kayts Island, Elephant Pass and Palali camps while the troops in Jaffna broke out of the fort in a column covered by several armoured vehicles. The operation met with stiff resistance, none of the columns being able to move more than a few kilometres. The column from Elephant Pass, said to number about 1000 with around 40 armoured and other vehicles was brought to a halt at Iyakachchi on the Jaffna–Kandy Road. According to some accounts the army tried to break out of this blockade for two days but failed. The militants are also said to have blasted the bridge at Pallai further north on the Jaffna-Kandy road.

---

[53] Interview with Brigadier. Bahar Morseth; Shamindra Ferdinando, 'Sri Lanka: the War on Terror Revisited; Sri Lankans Conduct Live Firing Exercises ahead of operation Liberation', *The Island*, 20. 6. 13, http://slwaronterror.blogspot.com.au/2013/06/israelis-conduct-live-firing-exercises.html

The Jaffna fort column left in four armoured vehicles towards the islands but returned to base after meeting with heavy resistance. The militants opened fire from both ends of the Pannai causeway bringing the vehicles to a halt in the middle of the causeway.[54] A potentially disastrous situation was looming when covering fire from the fort and a helicopter gunship enabled them to return safely. Closer to the fort, several hundred soldiers occupied the police married quarters next to the fort. The column from Velvetithurai met with a similar fate. An army detachment trying to move out of the camp on foot was beaten back by militants. The foray from Kayts also had to return after a brief probe. The detachment from Palali had some success, managing to reach Vasavilan Central College where they set up a new camp[55].

On 22nd May the army launched the second phase of Short Shrift, aimed at capturing the Mandaitivu island to the west of the Jaffna fort. This was an important move as Air Force flights landing supplies for the beleaguered Jaffna garrison came under fire from militants based on the island. The island also functioned as a base from which the militants could lob mortars and fire rockets in to the fort. The operation was successfully completed, relieving pressure on the garrison.[56]

The operations raised hopes in the South of a general push to retake Jaffna. Whether Short Shrift I and II were ever intended

---

[54] Pannai causeway links the Mandaitivu island with the Jaffna peninsula. It is about two miles long.

[55] *Sri Lanka Situation Report*, Tamil Information and Research Unit, Madras, no. 14 (1. 7. 86), pp. 4-5, Mervyn de Silva, 'Operation Turnaround or Turnabout?', *Lanka Guardian*, 1. 6. 86, p. 3, O'Ballance, *Cyanide War*, p. 16.

[56] L. M. H. Mendis, *Assignment Peace, in the name of the Motherland*, Author publication: Nugegoda, Sri Lanka, 2009, pp. 50-52.

as the beginning of such a general offensive or a precursor to one is a moot point but what is clear is that their achievements were modest. In a candid assessment the Defence Ministry spokesman admitted that the troops advancing from the different camps had failed to link up.[57] But Major-General Nalin Seneviratne, usually a sober and forthright appraiser, asserted that the army withdrew because it had met its objectives of securing the airfield and naval base and keeping them free from attack. In this respect they were somewhat successful. On the mainland troops now occupied Vasavilan Central College where a small camp was set up. When on 22nd May, 1986 Operation Short Shrift II succeeded in bringing the island of Mandaitivu under control the western flank of the Jaffna garrison was also secured.[58] The capture of Mandaitivu also enabled the Air Force to land supplies for Jaffna without too much hindrance from the sniping guerrillas. The security forces admitted to few casualties in these operations, no more than a handful killed and wounded. They however, claimed to have killed a large number of guerrillas, always a difficult claim to verify. However, dozens of people were confirmed admitted to the Jaffna hospital including two hospital staff struck by bullets fired from helicopters.[59] In Velvetithurai numerous houses were reported damaged by shelling by naval craft and due to aerial bombing.[60]

The success, however, was short lived. The operations did nothing to stop further attacks on the camps. Within days of completing the operation the army was still struggling to send out foot patrols and defend their bases against mortar fire. Heavily fortified strong points had reappeared within

---

[57] 'Security Operations Come to a Halt', *The Island*, 21. 5. 86, p. 1.
[58] 'Sri Lanka: the Siege Within', *India Today*, 15. 6. 89, p. 121,
O'Ballance, *Cyanide War*, p. 63, Mendis, *Assignment Peace*, p. 52.
[59] 'On the northern Front (Reuters Despatches)', *Lanka Guardian*, 1 .6. 1986, p. 4.
[60] 'VVT the Facts', *Saturday Review*, 31. 5. 86, p. 8.

site of the camps. One such strongpoint in Velvetithurai was located in a house barely 500 meters from the military camp and was defended by numerous pill boxes and trenches.[61] Nothing much seems to have changed except increasing the targets for guerrilla attacks.

In the Vanni too, the security forces made forays, and again, with mixed results. Clearing the Vanni of the rebel presence had been a daunting prospect for the armed forces, already stretched to their limit by troop commitments in Jaffna and the East. Finally in October 1986 the Army made an attempt to utilise their limited means effectively to strike the Tiger camps in the hinterland of Mannar. The focus was on the so-called Mannar rice bowl, the fertile area on the mainland to the southwest of Mannar island. Under the leadership of Denzil Kobbekaduwa the Army made a two-pronged approach, from the southeast and the northwest. The thrust from the southeast was a feint to draw the rebels in while the main thrust from the northwest cut off their communications with the coast. The rebels fell for the ruse and in the ensuing fighting suffered heavily. One of the Tigers killed was Victor, the man behind the Anuradhapura attack in May 1985. At least 20 other rebels died, among them several female cadres who were receiving their baptism of fire in this campaign. The security forces too suffered grievously. Fifteen soldiers and one Air Force crew member died in the fighting while at least another fifteen troops were reported wounded. Two soldiers also fell into the hands of the enemy for the first time. The bodies of nine of the dead soldiers were also captured by the tigers and taken to Jaffna where they were handed over to the Army in the Fort after being paraded before the public.[62]

---

[61] Smith, *Reluctant Mercenary*, pp. 146-7.

[62] 'Leading Tiger killed in Adampan', *The Island International*, 22.10.86, p. 1, Adele Balasingham, *Women Fighters of Liberation Tigers*, pp. 17-35,

The Adampan campaign, as it came to be known, was a cleverly conceived and executed operation making optimum use of scarce resources in a campaign of manoeuvre. Still, it was not sufficient to break the hold of the rebels in the western Vanni. The Tigers were back in business in the Adampan area by the end of the year and the Army back in their camps. Moreover, the operation also demonstrated the control the rebels had over the region. They were able to transport the bodies of the slain soldiers all the way to Jaffna by road in perfect security.

4.3 The Reluctant Mercenary

Tim Smith retired from the British army as a Warrant Officer II and a qualified helicopter instructor, having 'amassed a grand total of two ex-wives, half a house, half a car and half a golden handshake.'[63] He drifted from one flying instructor job to another until he received a telephone call from a long-lost friend who claimed he was training helicopter pilots for the Sri Lankan Air Force. The friend directed him to the company that recruited such instructors and Smith, unemployed and bored, was only too glad to avail himself of the opportunity to earn a tax-free income in an exotic part of the world he had heard little of.

Smith's employer was Keeny Meeny Services (KMS), the shadowy company that employed ex-servicemen as mercenaries. As explained earlier KMS had begun its engagement with Sri Lanka in 1984, initially providing training to the island's Special Task Force. Now they were branching out into offering their services as flying instructors for the SLAF. Smith was one of several foreigners who would end up spending many months

---

'Mannar Battle Victims Buried', *The Island International*, 22.10.86, p. 1,
'Truly a Hero of Our Times', http://www.slnewsonline.net/kobbe.htm
[63] Smith, *Reluctant Mercenary*, pp. 4-5.

training Sri Lankan helicopter pilots and even flying helicopters on combat and rescue missions.

Smith's experience in Sri Lanka, which he recorded in his memoirs *Reluctant Mercenary*, provides fascinating insights into the struggles of a cash-strapped and inexperienced Sri Lankan military against a robust insurgency. At the time of Smith's arrival, in mid 1986, the siege of the military camps is in full swing and the soldiers penned up within the barbed wire of their bases. As soon as Smith arrives, Greg, one of his colleagues, fills him in on the plight of the local army. A large military force had ventured about 10km out of Elephant Pass only to be ground to a halt by 22 culvert bombs and landmines. Many of the soldiers, Greg reveals, had "just sat on the roadside crying and refused to go another step further."[64] Such failures, we soon learn from Smith himself, are what contribute to the siege situation, the army failing to move beyond the perimeters of their camps. The camps themselves are under constant mortar attacks, the descent of shells on some camps almost akin to a nightly ritual: 'Everyday at about half past nine or maybe ten o'clock, the Liberation Tigers of Tamil Ealam gave the army a nightly salute.'[65] They seem to have caused little damage or casualties except for the odd mortar that hit its mark.

Smith's memoirs are particularly revealing in the way the Sri Lankan Air Force, the arm which Smith served, operated under difficult and challenging conditions. Smith arrived in Sri Lanka at a time when the Air Force had acquired new aircraft and heavier firepower in the form of additional helicopters and the Siai Marchetti ground attack aircrafts. The army's predicament placed an enormous burden on the expanded Air Force, especially its small fleet of helicopters which were required to fly numerous missions to supply and support the beleaguered

---

[64] Ibid., pp. 31-2.
[65] Ibid., pp. 55-6.

garrisons. Predictably, the first reaction to a sustained attack on a garrison was to get a helicopter in the air and once in the air the effectiveness of the machine was governed by the poor quality of equipment as well as the quality and temperament of the crew. The rockets lacked accuracy because they were not properly cleaned and the ammunition was of such a cheap variety that 'when fired at a target 200 metres away the rounds were hitting the ground somewhere from 80 to 300 metres away and to 20 metres to the left or right.' [66]

The personalities of the crew seem to determine the kind of response from the helicopter pointing to a lack of standard operating procedures. This does not augur well for the security forces or the civilians who are usually caught in the cross-fires. Namal in particular is a serious and serial offender. During one of the sorties Namal points out a man on a bicycle telling Smith that the way to figure out if he is a terrorist is to shoot and see if he runs. Namal shoots and the man falls dead. On another occasion during operations in Trincomalee Namal blasts away at a civilian taking refuge in a tree, cutting him to shreds. His gung-ho attitude also delivers a severe blow to the Navatkuli causeway, blowing almost half of it up in an attack on a Tiger vehicle. [67]

Such behaviour appears to be confined to Namal but it also points to a culture that turned a blind eye to it, perhaps a result of a modest establishment creaking under the pressure placed on it by the burgeoning military conflict. Still, despite the poor equipment and the lack of discipline the Air Force is able to deliver some telling blows when the enemy presents a target. During an abortive Tiger attack on a vehicle checkpoint at Thondamanaru and an army assault on a rebel stronghold in

---

[66] Ibid., pp. 67-8.
[67] Ibid., p. 85.

Velvetithurai the helicopters catch the enemy in the open and mow them down by the dozen.[68]

Smith's writing is peppered with caustic observations of the Sri Lankan security forces ranging from his condemnation of the Officer Corps of the Sri Lanka Army as a 'bunch of selfish bastards' who knew little or nothing of order and organisation and displayed a 'lack of control over their own actions and behaviour, and seems to have little interest in or control over their actions'[69] to his conclusion that the Sinhalese are a race without any aptitude for war or work. Such observations are more likely the result of his ignorance of the context of the war rather than any innate superiority-complex. Smith, one realises, is utterly ignorant of the military and political circumstances of the war. The Sri Lankan security forces in 1986 were far from professional outfits despite their organisation along British lines. And unlike the British Army, to which Smith had belonged, it had no long tradition and experience as a professional body. In this sense Smith was probably correct in describing the Sri Lankan Army officers as a 'grotesque parody of British officers,' wearing their uniforms and adopting their attitudes without the professionalism that went with it. Such empty posturing was fine as long as there was peace but with the escalation of the conflict, the brittleness of the foundations was exposed. It is for the revelation of the way that this lack of experience was displayed rather than the reasons for Smith's perceptions that his observations are important.

Smith however is largely confined by his own operational dimensions to be a witness to the wider air war that was being waged. His services were limited to the operations of the helicopters. The overworked Bell 212s and Bell 412s were certainly playing a big role in the conflict at this stage but the

---

[68] Ibid., pp. 80-2.
[69] Ibid., p. 72.

induction of the fixed wing ground-attack aircraft was also beginning to have a significant impact on the way the war was being conducted from the air. The Air Force used its newly acquired Siai Marchetti jet trainers for the first time in a bombing attack in Kilinochchi in February 1986. A large force of guerrillas had surrounded Kilinochchi army camp and the light aircraft were used to relieve pressure on the garrison.[70] A few days later the Siai Marchettis conducted their first bombing raid on Jaffna on 19 February 1986, when five aircrafts carried out an attack on what was a suspected Tiger camp that lasted nearly one hour. But the rockets fired and bombs dropped by the Siai Marchettis were no more accurate than those unleashed by the helicopters, causing more harm to the civilians than the insurgents. The trend was set from the very first bombing raid on Jaffna on February 19. Dubbed "the first aerial attack in Sri Lanka since WW II" by the Jaffna based *Saturday Review*, the bombs reportedly killed seven civilians and damaged several houses and a rice mill.[71] The following week the same area was bombed again, this time by three planes supported by two helicopters. Three civilians were reported killed and five injured[72]. It was said derisively that the SLAF tried to bomb a guerrilla camp in a crowded market area four times during April and May 1987 and failed to hit the target each time, instead killing and wounding dozens of civilians.[73]

This was a dangerous way to carry out counterinsurgency warfare, especially in Jaffna which was essentially a built-up urban environment. Tigers operated in small groups in this urban setting, presenting very few discernible targets for the Air Force. The weapons in the aircrafts were also not the most accurate. The result was that the bombing and strafing by the

---

[70] O'Ballance, *Cyanide War*, p. 56.
[71] 'Jaffna Bombed', *Saturday Review*", 22. 2. 86, p. 1
[72] 'Death rains from the Skies', *Saturday Review*, 1. 3. 86, p. 1.
[73] Hoole et al, *Broken Palmyrah*, p.106

helicopters and the light aircraft added to the woes of the civilians, joining the artillery fire from the camps. Scores were killed or wounded while leaving whole swathes of Jaffna in ruins. Just as the buildings in the fort were getting gradually pulverised by the militants the area immediately around it was also getting worked over by the air force and the artillery. Lalith Athulathmudali himself admitted that the bombings have been inaccurate and suspended air attacks in March 1986.[74] But the perceived advantage afforded by the new weapons was too great for the government to resist for the sake of civilian casualties. Air strikes resumed later that month bringing more death and destruction on the city.[75]

However, the increased firepower and air attacks did have some positive outcomes for the military. It boosted the morale of the soldiers. A northern journalist noted wryly how the soldiers stranded on the Pannai causeway during Short Shrift II waited for the helicopters and the cannon from the Fort to drive the Tigers away without firing all around nervously[76]. And as Smith reveals, sometimes, when the guerrillas could be caught in the open the aircraft could give them a terrible mauling as a group of guerrillas found during an attack on a military outpost at Thondamanaru in June 1986. But such successes were too few and far between. What happened more often was that the rockets, bombs and bullets hit civilian dwellings, causing casualties.

But the Air Force was also a problem for the Tigers. It may have killed civilians but it also curtailed the guerrillas' movements and kept the army garrisons from being completely cut off. Anton Balasingham admitted as much when he said in

---

[74] Michael Hamlyn, 'Tamils Step up Attacks on Civilians: the Communal Conflict in Sri Lanka', *The Times*, 11. 3. 86.
[75] Humphrey Hawkesley, 'Sri Lanka Resumes Air Raids on Tamil Villages', *The Age*, 29. 3. 86. The first raid upon resumption on villages near Palali reportedly killed one man and injured 15.
[76] 'Victory and Defeat', *Saturday Review*, 24. 6. 86, p. 2

March 1986 that, if they could neutralise the government's air superiority the balance will be in the Tigers' favour.[77] There were reports of attempts to acquire SAM missiles to counter the air threat but these had not been successful. But it did not prevent the militants from unleashing what they had on the Air Forces helicopters and bombers. During the early days the guerrillas had simply let loose their AK 47s at the air craft whenever they were within range. Later with the acquisition of several .50 calibre machine guns this changed. These guns were usually mounted on pick-up trucks and moved from place to place as the situation demanded. Fear of these guns made the helicopter pilots reluctant to fly over Jaffna city at an altitude less than 3000 feet.[78] It was a well-founded fear. In March 1987 the fire from these guns severely disabled one helicopter during an operation in Kattuvan near Palali. Another helicopter suffered serious damage from ground fire while providing support for troops in the Velvetithurai camp in April 1987.[79] In another incident around this time the fire from .50 calibre gun caused a fire in one of the engines of a Y-12 aircraft coming to land in Palali[80]. The aircraft landed safely and was made airworthy again later but it showed that the sky was not totally out of limits to the Tigers. At times even RPGs were fired at helicopters. In one bizarre incident an RPG with the safety pin on crashed into a helicopter hitting the arm of civilian in the aircraft and cutting the fuel to one of the engines.[81] This vulnerability of the Air Force eased the pressure on the rebels to an extent as now the helicopters had to stay at a reasonable distance from their targets.

---

[77] Humphrey Hawkesley, 'Tigers Find Teeth For a War at Sea', *The Age*, 5. 3. 86.

[78] Smith, *Reluctant Mercenary*, p. 50.

[79] Jagath P. Senaratne, *Sri Lanka Airforce: a Historical Retrospect*, 1985-1987, (Colombo: Sri Lanka Air Force, 1998), volume 2, p. 79.

[80] Athas, 'Rendezvous in Madras to Receive seized Arms', *Weekend*, p. 7.

[81] Senaratne, *Sri Lanka Airforce: a Historical Retrospect*, 1985-1987, volume 2I, pp. 74-5.

Enhanced firepower and command of the air helped prevent the Tigers from pressing their advantage home but it did little to extend the military's control over the north. The army remained cooped up in their camps while the militants roamed around freely. Even in their bases the soldiers were not free from guerrilla harassment. For their part, the guerrillas were able to keep the Army inside the bases but lacked the firepower to pound them to submission or to overrun them.

4.4   The Naval Challenge

In the sea the naval surveillance zone worked, but only up to a point. The Tigers were using boats that were between 15-20 feet long and which showed little freeboard when loaded. They were usually fitted with three outboard motors, 'two for getting across the [Palk] Straits, one for loitering before they beached, and three for running like hell is they were spotted by the Navy's Coastal Fast Patrol Boats" (FPBs) or the Air Forces Skymaster Offshore Patrol plane."[82] They were unarmed and depended on stealth and speed to evade the navy. The coastal stretch to the north of Trincomalee Bay was a favourite area for unloading supplies. The rebel boats plying the coastal waters hugged the coast to avoid detection and if spotted by the Navy they beached their boats and slipped inland.[83]

But the Tigers were also exploring ways of combating the navy. In early March 1986 a naval boat was blown up by a limpet mine panted by the guerrillas, killing two sailors and wounding three. Ominously for the navy, the guerrillas also announced that they were getting ready a team of 'Sea Tigers'.[84]

---

[82] Smith, *Reluctant Mercenary*, p. 129.
[83] Iqbal Athas, 'The Stinging Bee Experience', *Weekend*, 14.9.86, p. 6 and p. 10.
[84] Humphrey Hawkesley, 'Tigers Find Teeth for a War at Sea'. Tamil sources claimed to have killed 11 naval ratings and one officer in the

Like the army and the air force, the navy had also gained much hardware since 1983. The Dvoras they soon found, were capable of matching the fastest Tiger crafts in speed. The rebels were becoming increasingly wary of these new fast boats. One Tiger captured by the navy said that his boat was twice deterred by Dvoras from entering Sri Lankan waters off Trincomalee.[85] This was far from sufficient to keep the traffic of arms and material in check, but there were more frequent interceptions now. On 27th April 1986 the navy scored a major victory when they confronted a fibreglass boat carrying Tiger cadres. In the ensuing battle the Tiger boat was destroyed killing six of the occupants. The dead included the Eastern Province commander of the LTTE Captain Aruna. The navy also recovered six G3 rifles and two Browning heavy machine guns. Three days prior to this the navy had also intercepted another boat belonging to the TELO killing 10 guerrillas.[86]

But such interceptions were not sufficient to halt the flow of arms and men. The LTTE was now also relying less on India than on its own shipping arrangements. The Tigers' merchant ships off-loaded their military supplies into small boats in the Bay of Bengal and the boats ferried them to Sri Lankan shores. According to military intelligence the rebels depended on India only for explosives at this stage.[87]

---

attack. *Sri Lanka Situation Report*, Tamil information and Research Unit, Madras, no. 9, (15. 4. 1986), p. 3.
[85] On the third occasion the boat and another were intercepted by Dvoras. Iqbal Athas, 'The Stinging Bee Experience.'
[86] *Sri Lanka Situation Report*, Tamil Information and Research Unit, Madras, no. 13, (15. 6. 86), p. 3.
[87] Michael Hamlyn, 'Tamil Attacks Harden Attitudes: Sri Lankan Despair over Ethnic Crisis', *The Times*, 22. 4. 87.

CHAPTER 5

Breaking the Shackles - 'Giant Step' and 'Liberation'

The second half of 1986 was a time of intense activity on the peace front. Rebuked by India after the Short Shrift operations and rebuffed by the Tigers in Jaffna, the government finally began working on a set of proposals to end the conflict peacefully. The proposed reforms envisaged the devolution of power to the provinces. And like previous attempts at peace it was doomed from the start.

5.1   The Elusive Peace

For the first time since the conflict broke out, Jayawardena seemed serious about offering a political solution. The main problem for the president was getting the endorsement of the Sinhalese South which was very susceptible to the influence of the Buddhist clergy. Jayawardena succeeded in getting the unanimous endorsement of his party's national executive committee, but getting the clergy and the opposition on board was a different matter. The proposed reforms were proving a stimulus to the long dormant Opposition and they were not losing any time in exploiting it. The leader of the main opposition party in the south, Mrs. Sirimavo Bandaranaike rejected provincial councils as being useless in helping any community and went to see the Buddhist prelates to canvass their opposition.[1] Spearheaded by the SLFP and the Buddhist clergy, the Opposition coalesced into the Movement for the Defence of the Nation (MDN), drawing together a wide array of organisations and individuals. The MDN began holding meetings around the country, mobilising support against the proposals.

While Jayawardena struggled to get the south on board, India struggled with swinging the Tamil opinion in favour of the proposals. The problem for India was that they were fast losing control over the militants. If the manner of the Tiger's eliminating of the TELO was brutal, the timing of the massacre was crucial. On May 3rd two special envoys of the Indian Prime Minister, P. Chidambaram and K. Natwar Singh were due to visit Colombo to discuss ways of moving negotiations forward. The elimination of the TELO, the group closest to India, sent a powerful message about the changing ground reality in the North. The Tigers were on top and India now had to reckon with their supremacy when dealing with the Sri Lankan government. The TULF still commanded respect among educated, moderate Tamils but they had very little, if any, authority over the events in the north. They could only take their cue from the Tigers.

The Tigers - and the TULF - rejected the government's proposals. The sticking point was the amalgamation of the Eastern and Northern provinces into one unit. The Tigers emphasised that "for any meaningful political settlement, the acceptance by the Sri Lankan government of an indivisible single region as the homeland of the Tamils is basic."[2] Whilst such pronouncements served to strengthen

---

[1] "'No Community will Benefit from Provincial Councils'"- Sirima', *Lanka Guardian*, 15.8.86, p. 8.
[2] P. Venkateshwar Rao, 'Ethnic Conflict in Sri Lanka: India's Role and Perception,' Asian Survey, Vol. 28, No. 4 (April., 1988), p. 429.

the position of the hardliners in the south who saw the militants as inflexible, Jayawardena's attempts to placate the southern constituency did little to help ease mounting tensions. Jayawardena hit out at the militants warning that no legislation to devolve power will be enacted until the militants laid down their weapons.[3] Such declarations played into the hands of the LTTE who saw any solution offered by the Sinhala leadership as eyewash. It also did nothing to endear the president to the southern pubic who judged him for his attempts to devolve power in the first place. It certainly didn't intimidate the Tigers.

India too felt increasingly exasperated by the Tigers' behaviour. And while Jayawardena battled southern intransigence India decided to turn up the heat on the Tigers. After the rebels had fired into a crowd in Tamil Nadu on Diwali Day in November 1986, the Indian government came down heavily on the Tigers. On 8th November they raided the hideouts of five guerrilla groups in a state-wide crackdown, arresting militant leaders and confiscating a massive haul of weapons. Among the many arms and equipment captured were thousands of AK47, RPGs and, alarmingly, several SAM-7 missiles.[4] The Tiger leaders were later released but the weapons were lost forever. India was sending the militants a clear message: if you want to enjoy our hospitality, play by our rules.

Undaunted, the Tigers transferred whatever weapons that had eluded the Indians to northern Sri Lanka. In fact the transfer

---

[3] 'Tactical Offensive on the Diplomatic Front or Tactical Retreat on the Domestic?', *Lanka Guardian*, 15. 10. 86, p. 3

[4] S. H. Venkatramani, 'Taming the Tigers', *India Today*, 30.11.1986, pp.22-3, 'India Cracks the Whip', *Asiaweek*, 23.11.86, p. 33. The offices belonged to the LTTE, PLOT, EPRLF and EROS. The confiscated weapons and equipment was valued at R.s 40 crore. 'Operation Tiger: the Reasons Behind', *The Island* 7.12.86, p. 9 (courtesy *Amrita Bazaar Patrika*).

was already underway when the Indians struck. Alarmed by the growing dissatisfaction of the Indian administration with their stance towards peace talks the LTTE had started shifting their bases to Jaffna earlier that year.[5] Prabhakaran himself returned to Jaffna in January 1987, after a sojourn of several years in India. The Tigers, it seemed, were making preparations for establishing a separate rebel administration in Jaffna as a step towards a declaration of independence. As a preliminary to this, in the first week of January 1987 they announced their intention to issue motor vehicle licences. An incensed Athulathmudali promptly declared a fuel embargo on Jaffna. Petrol bowsers were turned back at the Elephant Pass camp. The minister also threatened to cut off electricity to the peninsula. Within days fuel prices skyrocketed in Jaffna, a litre of petrol which cost Rs 17.50 before the embargo now spiralling up to Rs.50. Along with it rose the price of commodities.[6] Transport came to a standstill resulting in food shortages while people stayed away from work and garbage piled up on the streets.[7] The consequences for the residents in Jaffna were so dire that the usually combative *Saturday Review* made an uncharacteristic "humble appeal" to the militant groups to rethink their strategy "at this hour of great peril to the Tamil people."[8]

5.2   Giant Step

With the peace process at a standstill and the rebels seemingly determined to establish their separate state, the stage was set for another military confrontation. It came in February 1987 when the government launched its most ambitious military operation yet. Code named "Giant Step,"

---

[5] Jon Swain, 'Face to Face with the Guerrilla Commander: Cyanide Martyrs Bar Way to Peace', *Sunday Times*, 10. 8. 86, reproduced in *Lanka Guardian*, 1. 9. 86, p. 11.

[6] 'Bearing the Blockade', *India Today*, 15. 2. 87, p. 27.

[7] 'Grain Cannot be Moved into Jaffna', *Lanka Guardian*, 1. 2. 1987, p. 12.

[8] 'Economic Sanctions Against Jaffna', *Saturday Review*, 10. 1 .87, p. 12.

the operation was launched simultaneously in the districts of Jaffna, Mannar, Kilinochchi, Trincomalee and Batticaloa in early February and continued intermittently till the end of March. The objective was similar to Short Shrift 1 and 2: to clear areas around major military encampments and establish new outposts. Another major objective was to dominate areas on the north western and north eastern coasts and deny these areas to the guerrillas who were using them to insert men and material into the Vanni. It was the first time the military had launched an operation on such a scale across the north and East, and as such, it also provided a great opportunity to give the troops a taste of bigger things to come.

The security forces reported the operation as a success claiming to have cleared large areas around their camps. In Mannar troops moved south east towards Uyilankulam and northwards towards Adampan and then towards Illuppukadavai further north. In Trincomalee, areas around Kurumbupiddy up to Pulmoddai and Yan Oya was reported completely cleared. A massive base camp was captured at Periyakarachchi to the north of Illupukadavai. In kilinochchi two military columns from Kilinochchi and Elephant Pass converged on Paranthan, securing the town and setting up road blocks. Troops occupied government buildings in town, including the post office and the chemicals factory. In Batticaloa troops ventured out of camps in Kokkadicholai and Vavunativu and destroyed several militant hideouts. In Jaffna too, troops made forays into rebel territory. They thrust out from Palali and established control over the Thondamanaru bridge while a large minefield of over 400 yards was reported cleared outside the Vasavilan camp. They also ventured out of the fort and captured the Jaffna telecommunications exchange and the buildings in the vicinity. These buildings were taken to deny the Tigers vantage points from which to harass the soldiers in the fort[9]. Another aim of the operation was to cut off land communications between Jaffna town and the countryside but this was far from successful. The fact that the advance was measured in meters rather than miles gave a good indication of the challenge faced by the troops.[10]

In March after constant sniping and mortar attacks, the troops again moved out of Palali to clear the area around the Kattuwan junction to the south of the Palali air base and established an outpost. An area of approximately four miles around the junction was declared cleared by the troops. The troops also moved out of the Navatkuli camp to the Navatkuli junction to extend its perimeters after the Tigers pounded the camp with mortars for nearly 20 hours on 7th and 8th March.[11]

Although the Armed Forces reported success in the operation the gains were clearly limited. Dozens of "terrorists" were reported killed and captured but the veracity of such reports was always suspect. The only serious fighting seemed to have occurred around Uruthapuram north east of Kilinochchi where the military column from Elephant Pass ran into some resistance, suffering several killed and wounded. However, the items reportedly captured from the guerrilla hideouts painted a dismal picture. It was mostly equipment and material like vehicles, stocks of fuel which were difficult to move in a hurry that fell into the hands of the security forces, an

---

[9] 'Security Forces Secure Large Areas', *Sunday Observer*, 15. 02. 87, p. 1, Mendis, *Assignment Peace*, pp. 57-8.
[10] Edgar O'Ballance, *Cyanide War: the Tamil Insurrection in Sri Lanka 1973-88* London: Brassey's, 1989, p. 73-5, Iqbal Athas, "Operation Giant Step", *Weekend*, 22. 2. 87, p. 7.
[11] Iqbal Athas, "Rendezvous in Madras to Receive Seized Arms", *Weekend*, p. 7, Iqbal Athas, 'Peace through the Ballot?', *Weekend*, 15. 3. 87, p. 7.

indication that the guerrillas had retreated before the oncoming troops.[12] This was only to be expected. The guerrillas did not have the strength to meet a strong security forces contingent head on except perhaps in the Jaffna peninsula. Besides, the security forces gave the guerrillas plenty of warning about their designs. The fledgling military struggled with its transport which relied heavily on the few transport aircraft and road bound vehicles. This meant that a troop build up took a considerable time giving sufficient warning to the enemy of the impending operation. This probably explains the unusually light mortar fire experienced by the army during its advances.[13]

The operation also piled more hardships on civilians already smarting from the spiral of violence. Tamil sources claimed that 20 civilians were killed and 93 injured during the operations in Kilinochchi while 3000 people were displaced in the Paranthan area.[14] These sources also alleged that Adampan hospital was bombed on 11th February killing several patients.[15]

The new wave of violence elicited a heated response from across the Palk Straits. The crackdown on the LTTE in India had not meant that the Indian government was empathising with its Sri Lankan counterpart. Indeed India had warned soon after the raids in Tamil Nadu that any attempt to launch fresh offensives against the Tigers in the wake of the raids would undermine India's attempts to find a solution to the national question in Sri Lanka.[16] Clearly

---

[12] L. M. H. Mendis, *Assignment Peace, in the Name of the Motherland*, Author publication: Nugegoda, Sri Lanka, 2009, pp. 57-8,

[13] 'Tigers at Bay', *The Economist*, 28 .2. 87, p. 28.

[14] 'Kilinochchi Faces Famine', *Saturday Review*, 7. 3. 87, p. 8.

[15] 'Adampan Hospital Bombed', *Saturday Review*, 14. 2. 87, p. 8.

[16] John Elliot, 'India Warns Sri Lanka of Offensive', *Lanka Guardian*, 1. 12. 86, p. 3.

India's perception of Sri Lanka's commitment to the Peace Process was not very rosy. Now, with the Sri Lankan security forces apparently launching a major offensive, Prime Minister Rajiv Gandhi expressed his displeasure in no uncertain terms, announcing that India was suspending its mediation. Military action will only make the situation escalate, he warned. [17] The Sri Lankan government took an equally tough stance. There was to be a cessation of hostilities only when the militants had laid down arms.[18] The government did suspend air strikes after the Indian rebuke but they were resumed in March along with heavy shelling from the Jaffna fort that reportedly killed 17 civilians and injured dozens more.[19]

It was now clear that a showdown was imminent. Continuing hiccups with the Peace Process seemed to have convinced the Sri Lankan government that the military option was the best and perhaps the only option. To this end, they were willing to test India's resolve or were expecting Indian frustration with the Tigers to translate into empathy for the Sri Lankan state in the case of a military offensive. The Tigers, it seemed, were willing to test the resolve of both India and Sri Lanka, with the latter to reveal its military intentions and with the former to persuade them to throw in their lot more closely with the rebels.

And the recent operations hinted at the government's future military strategy. The security forces were going to consolidate

---

[17] P. Venkateshwar Rao, 'Ethnic Conflict in Sri Lanka: India's Role and Perception', pp. 431-2.

[18] A. J. Wilson, *Sri Lankan Tamil Nationalism: Its Origins and Development in the 19th and 20th Centuries*, London: C. Hurst and co., 2000, p. 150.

[19] Rajan Hoole et al., *Broken Palmyrah: The Tamil Crisis in Sri Lanka, an Inside Account*, Sri Lanka Studies Institute: Ratmalana, Colombo, 1992, p. 101.

their power in the East and the Vanni before embarking on any major operations in Jaffna. The Giant Step operations also gave an indication of the LTTE's likely military strategy in the event of a major incursion into their territory. The rebels had offered little resistance to the security forces outside the Jaffna peninsula. In fact, they withdrew most of their cadres and leaders from other areas, including Mahaththaya from Vavuniya and Radha from Mannar into Jaffna.[20] In Jaffna, they fought hard to slow the army's advance. Once the army began consolidating their meagre gains the Tigers also counter attacked. Early in the morning of 23rd March they launched a sustained rocket and mortar barrage on the Jaffna fort and its outposts. While the garrison in the fort was pinned down by the mortar and rocket fire a group of Tigers stormed the outpost at the entrance to the Pannai causeway where six soldiers and three policemen were stationed. After a brief fire-fight that killed a militant and a soldier the rebels surrounded the outpost. A party sent to reinforce the outpost was ambushed, killing a sergeant and three soldiers. In the meantime, with their ammunition exhausted, the remaining soldiers and policemen at the outpost surrendered to the Tigers.[21]

The attack, especially the capture of soldiers, was a blow to the military's morale that had risen after the recent operations. It seemed that the Tigers were always finding a way to bounce back. However, the rebels were not always successful with their counter attacks. On 14th February, a week into Operation Giant Step, a major attempt to overrun a military camp in Jaffna backfired when an accidental explosion destroyed the booby trapped vehicle that was to be used in the attack. The vehicle, a water bowser that transported water to the Navatkuli army camp, was

---

[20] Ibid., p. 99.
[21] Iqbal Athas, 'Bid to Internationalise Hostage Crisis by Tigers', *Weekend*, 29. 3. 87, p. 7.

provided with a false bottom that concealed explosives to be set off by a booby trapped device when the bowser was inside the camp. The Tigers were to storm the camp from several directions led by sandbagged trucks and under the cover of mortar fire. Unfortunately for the Tigers, a short circuit set off the explosives while the finishing touches were being laid in a rebel workshop, killing several of their leaders and dozens of civilians in the vicinity. The 300-400 soldiers in the Navatkuli camp were spared a terrible fate. [22]

The resistance to the military forays in Jaffna, the abortive attempt on Navatkuli, and the successful attack on the outpost on the Pannai causeway centre all showed the Tigers' commitment to seriously challenging the government's military power on the peninsula. It was a warning to the Sri Lankan military that taking Jaffna was a serious military challenge. They were better equipped and trained than they were a year ago but the Tigers were also stronger.

But even more disconcerting to the government was the LTTE's renewed assault on Sinhala villages in the East. On 7th February, the day Operation Giant Step got under way, in a calculated attempt to distract the security forces, they struck in Aranthalawa in Ampara killing 28 men, women and children, and almost all of them brutally hacked to death.[23] On 24th March, another village in the East, this time Serunuwara in Horowapotana, suffered the Tigers' wrath losing 26 villagers.[24]

---

[22] According to some reports 43 people were killed and 51 injured. Among those killed were 'Lieutenant Colonel Kugan, the Mullaitivu LTTE commander and 'Major Curdles', the Thenmarachchi leader. D. B. S. Jeyaraj, 'Ten Years Ago: The Kaithady Explosion', *The Island*, 16. 2. 97, p.16, Athas, 'Operation Giant Step', *Weekend*, 22. 2. 87, p. 7.

[23] H. W. Abeypala, 'Black Saturday's Slaughter House', *Weekend*, 15. 2. 87, p. 6 and p. 11.

[24] Elmo Perera, 'Bloodlust of the Brutal Tigers', *Weekend*, 20. 3 1987, P. 8 and p. 21.

In Jaffna the recent military operations were widely seen as a precursor to a large scale offensive to capture the peninsula. Tension ran high in the peninsula in anticipation of an imminent assault. Travellers to Jaffna reported soldiers at Elephant Pass check point telling them to 'keep their Indian planes ready' to make a quick exit because the soldiers were coming to take Jaffna soon[25]. In Jaffna Tigers were busy digging trenches and building bunkers. Buildings were being booby trapped and there were reports of bamboo stakes being driven to the ground in school playgrounds to counter landing by helicopter borne troops.[26] All the key LTTE leaders were called to Jaffna and the Tiger ranks in Jaffna were being reinforced with troops from other areas.[27] Jaffna streets became deserted by 6 pm as people sought the refuge of their homes. Many homes now had bunkers and many residents spent the majority of their time at home in these dinghy dugouts.[28]

Then in a sudden spirit of goodwill the government declared a ceasefire to cover the traditional Sri Lankan New Year period 11th -20th April. But the Tigers responded with two gruesome attacks. On Good Friday, 17th April, they stopped a convoy of civilian buses on their way to Trincomalee near the village of Aluth Oya on the Habarana-Trincomalee road and killed 127 of the passengers, many of whom were service personnel returning to duty after the New Year celebrations. Four days later a massive car bomb exploded in the heart of

---

[25] "We shall finish all our business elsewhere and come to meet the Tigers and you in Jaffna as soon as possible. Keep your Indian planes ready to make a quick getaway". 'See You in Jaffna', *Saturday Review*, 21. 2. 87, p. 8.
[26] Athas, 'Rendezvous in Madras to Receive Seized Arms", *Weekend*, p. 7.
[27] Hoole et al., *Broken Palmyrah*, p. 99.
[28] Ibid., pp. 114-15.

Colombo at the Pettah bus stand killing 110 people and injuring over 300. [29]

The massacres outraged the South and provided the government with the opportunity to launch the long awaited offensive. In Colombo President Jayawardena fumed. "We have had enough of their ceasing to use arms," he thundered. "These people have to be defeated militarily." He asked for money, arms and sympathy from the rest of the world. "Why don't you help us?" he pleaded.[30]

Jayawardena finally unleashed his troops on Jaffna in May. "Everybody told us, the government, you are doing nothing," explained Lalith Athulathmudali, giving reasons for the decision. "So we had to launch a crackdown on the terrorists."[31] The 'crackdown' came on 26th May with the launching of "Operation Liberation" the biggest military operation launched by Sri Lanka since Independence.

5.3 'Operation Liberation' - Objectives and Preparations

Vadamarachchi is the smallest and the least populated of the three geographical divisions of the Jaffna peninsula. Starting as a thin strip of land around Elephant Pass, it extends northwest, flaring into a broad, oblong area, making it appear like a smaller version of the peninsula. The narrow part of Vadamarachchi, known as the Vada strip, is sparsely populated while the top end is home to the vast majority of the inhabitants in the sector, concentrated

---

[29] O'Ballance, *Cyanide War*, p. 76.

[30] 'They Must be Defeated! Jayawardena Lambasts Tamil Rebels and India', *Time*, 11. 5. 87. p. 14. This was at a meeting with six Western journalists in early May, 1987.

[31] 'The Army Will Stay On', (Lalith Athulathmudali's interview with *India Today*'s Dilip Bob and S. V.Venkatramani), *Lanka Guardian*, 1. 7. 87, p. 9.

in an area roughly seven miles long and five miles wide. It boasts of two sizeable towns, Point Pedro and Velvetithurai, the latter the hometown of the Tiger supremo Prabhakaran and as explained earlier, a haven for smugglers at the time. There were also a few smaller population centres like Nelliady and Uduppiddy. Three main roads connected Vadamarachchi with the adjacent Thenmarachchi sector across the lagoon: the Palali – Point Pedro Road in the north that runs through Thondamanaru and Velvetithurai, the Jaffna-Point Pedro Road that crossed the lagoon at Achchuveli and further south, the road through Karaveddy along which one could travel to Chavakachcheri. Another road led form Point Pedro through Vallipuram all the way down the Vada strip until it joined the Sornapattu-Thalayadi Road which crossed into Thenmarachchi. In addition there were several minor roads in and out of Vadamarachchi. There was also the lagoon, not too deep for navigation by shallow craft or even to wade across. Last but not least, the coast skirted Vadamarachchi in the north and the east, offering another avenue of movement in and out of the region.

The objective of 'Operation Liberation' was to seal off and capture Vadamarachchi. The plan was to sweep the populated northern end of the area from west to east while the escape routes to the south and across the ocean were plugged.

Lalith Athulathmudali the minister of National Security, explained the focus on Vadamarachchi in terms of several strategic considerations. The area, he claimed, housed many of the ordnance factories of the Tigers and was also one of the main arms supply points.[32] It was also hoped that the military would be able to capture Prabhakaran. Military

---

[32] Ibid., p. 9.

intelligence had noted that the Tiger chief was in Velvetithurai at the time of the operation.[33] If they could net him, it would be a prize catch and a promise of a swift end to the war. However, according to Brigadier Gerry de Silva who was in overall command of Jaffna at the time, Vadamarachchi was chosen as the main theatre of operations because the president refused to allocate a brigade-size force to hold and dominate the Jaffna municipal limits. The military top brass had to settle for Vadamarachchi as the second best option.[34]

Whatever the reasoning behind the move on Vadamarachchi, the government seems to have envisaged a brutal campaign. Athulathmudali reportedly asked General Cyril Ranatunge to raze Vadamarachchi to the ground and even offered bulldozers to complete the job. The president expected Jaffna to be reduced to rubble by carpet bombing. [35] The desire to end 'the terrorist menace' once and for all was seemingly making the government incline towards a ruthless- and reckless- approach. The gloves were off and the government was not going to pull any punches.

For the security forces, the campaign came as a challenge as well as a relief. For the first time since the war began, and indeed for the first time in its short history, the Sri Lankan armed forces were embarking on a major combined-arms military operation involving several thousand troops aimed at capturing and holding territory. The experience of recent forays into rebel territory in the peninsula has shown that they could expect a stiff fight from a committed enemy. It was a major and daunting

---

[33] Hoole et al., *Broken Palmyrah*, p. 125.
[34] Gerry de Silva, *A most Noble Profession: Memories that Linger*, Colombo: International Book House, 2011, p. 75.
[35] Ibid., p. 75. Cyril Ranatunge, *Adventurous Journey*, Colombo: Vijitha Yapa, 2009, pp. 134-5.

challenge for the fledgling armed forces. On the other hand, the prospect of going on the offensive after months of languishing behind their sandbags with the aim of destroying rather than containing or pushing back the enemy came as a relief. Finally, it seemed, the war could be taken to the enemy and fought to a finish.

The Sri Lankan Armed Forces mustered a formidable force for the operation. The Air Force marshalled an impressive assembly of aircraft. Six Siai Marchettis functioned as ground attack aircrafts while two helicopters were deployed as gunships. One BAC-748 (AVRO), 2 Y12s, 1 De Haviland Heron were used as improvised bombers. Two BAC-748s, 2 Y12s and one De Haviland Dove were employed as transports ferrying goods and men. Eight helicopters also figured as troops carriers. Casualty evacuation also received high priority; one Y12 was devoted to that purpose.[36] It was still a poor man's Air Force but it was the biggest deployment of air power by Sri Lanka yet.

On the ground too, the preparations were no less impressive. At the Palali airbase, the launching pad for the attack on Vadamarachchi, the soldiers waited. They were organised in two brigades, the First and the Third, each about 1,500 strong. The First Brigade was commanded by Brigadier Denzil Kobbekaduwa while the Third Brigade came under the command of Colonel Vijaya Wimalaratne. Both were officers of distinction and considered competent leaders. The two brigades were to march up to Thondamanaru, about 4 km to the east of Palali and upon crossing the lagoon into Vadamarachchi, to move along two axes, sweeping Vadamarachchi from west to east while

---

[36] Jagath P. Senaratne, *Sri Lanka Air Force: a Historical Retrospect*, 1985-1987, Colombo: Sri Lanka Air Force, 1998, volume 2, p. 79.

the commandos blocked the exit routes. In the words of General Cyril Ranatunge, 'an imaginary line was mapped across Vadamarachchi area and Col. Vijaya Wimalaratne was entrusted with launching the operation from the northern theatre of Vadamarachchi. Brigadier Denzil Kobbekaduwa's brigade was scheduled to proceed north from south of Vadamarachchi strangling the LTTE and trapping them in the Vadamarachchi area.[37]

The First Brigade comprised the First battalion of the Gajaba Regiment and the First battalion of the Gemunu Watch, while the Third Brigade consisted of the 3rd battalion Sri Lanka Light Infantry (SLLI), 3rd battalion of the Gajaba Regiment with elements of the Engineers Regiment attached to them[38]. They were supported by artillery – mainly 120mm mortars and 76mm mountain guns and 85mm artillery from the Thondamanaru and Velvetithurai camps. The brigades also included details of armour – Saladins, Ferrets, Buffels, Saracens and also jeeps mounted with recoilless rifles – the armoured might of the fledgling Sri Lankan Army.[39] There was also an additional brigade, the Second, under the command of Brigadier Gerry de Silva. Its task was to create a diversion in Valikamam in and around Jaffna town and to plug the southern end of Vadamarachchi in the vicinity of Elephant Pass.

The operation was anticipated in Jaffna. As usual the build up had not been a secret and the Tigers had noted troops, arms and ammunition pouring in to Jaffna, to be landed at the Kankesanthurai port and then moved to Palali[40]. To

---

[37] Ranatunge, *Adventurous Journey*, p. 134.
[38] Mendis, *Assignment Peace*, pp. 62-3.
[39] Each brigade had a troop of armour - 1 Saladin, 1 Ferret and 2 Saracens. Jagath P. Senaratne, *Sri Lanka Armoured Corps: 60 Years' History*, Sri Lanka Armoured Corps, 2015, p. 117.
[40] 'Civilians? Hang Them!', *Saturday Review*, 23. 5. 87, p. 8.

keep the enemy guessing while they prepared for battle, the Sri Lanka military also engaged in diversionary tactics. According to a military source, empty helicopters flew back and forth between Palali and Point Pedro regularly in order to give the Tigers the impression that a heavy troop build up was taking place in Point Pedro for an operation from that quarter.[41] However, the rebels seem to have had their own ideas about the imminent operation. Expecting an all out assault on Jaffna town itself, the Tigers called the bulk of their cadres to strengthen defences in Valikamam where they were placed under Kittu and other senior commanders. Speaking in May 1990 Anton Balasingham ascribed this to misleading information provided by the Indian intelligence agency RAW. Questioning this, the analyst, the late Dharmaratnam Sivaram opines that the LTTE was unlikely to have been unaware of the military preparations to invade Vadamarachchi. For months, heavy road construction had also been taking place between Palali and Thondamanaru in preparation for the impending assault in that direction. Whilst Balasingham's assertion is questionable - though not entirely implausible - as it comes in the wake of the falling out between India and the Tigers after the Indo-Lanka Accord, Sivaram's view is not borne out by the concentration of the Tigers' main forces and commanders in Valikamam. The logistical preparations between Palali and Thondamanaru may have been interpreted by the rebels in the same way the helicopter flights seem to have been interpreted - as diversions, even though in this case with the potential for utility in the distant rather than immediate future.

The number of Tiger cadres preparing for battle in Vadamarachchi at this time is hard to estimate. The Tigers claimed in early 1987 that they had as many as 10,000

---

[41] Interview with Brigadier Bahar Morseth.

cadres but the government played down the numbers placing them at 1500-2000.[42] The commander of Jaffna Fort Lieutenant Colonel Asoka Jayasinghe believed that there were only about 300-400 Tigers in Jaffna city itself but that they could rely on additional cadres anytime.[43] The rebels no doubt inflated their numbers while the government downplayed them. The number of hardcore cadres was probably counted in their hundreds but there were certainly many more young men and women armed or willing to fight for their cause. The majority of these were certainly in Valikamam. The cadres were armed with a motley array of weapons. The heavy firepower was provided by a few .50 and .30 calibre machineguns, 40mm grenade launchers, RPGs and assorted mortars while the small arms ranged from the AK47 to shotguns supplemented by grenades including the primitive grenade launchers. How much of this modest firepower was deployed in Vadamarachchi is not clear but it is likely that the bulk of it was in Valikamam. But whatever their strength and wherever they were, the rebels now busied themselves strengthening their bunkers and setting up new defence posts. In Vadamarachchi itself, the area facing the Thondamanaru camp was liberally strewn with mines and defensive works such as bunkers and pill boxes set up. Everywhere in the peninsula people lived in dire anticipation.

5.4   In to Vadamarachchi

The security forces had begun probing attacks days before the main operation got underway. Air attacks on targets in the peninsula, especially in the Valikamam area, began immediately after the bomb blast in Pettah. From 18th May

---

[42] S. Venkatnarayan, 'A Visit to Jaffna', *The Island*, 15. 2. 87, p. 6.
[43] 'Tigers hold Troops as Captive Force', *Weekend*, 21. 6. 87, p. 11.

the army also made several forays from their bases in Palali, Thondamanaru, Kurumbasiddy, Kattuvan and Navatkuli. In one of these attacks Lt. Colonel Radha who was the LTTE's commander of the Jaffna District, was killed. The 30 year old veteran of many of LTTE's exploits in the past was one of the highest ranking Tigers to be killed by the Sri Lankan military until then.[44]

The operation proper began on 26th May. Despite the bravado of the political leaders about flattening Jaffna the military planners approached their task with a degree of sophistication. At dawn several helicopters belonging to the Sri Lanka Air Force arrived over the lagoon bringing commandos from the Special Forces, 'in a scene reminiscent of a movie' according to one officer.[45] Their task was to plug the exit points from Vadamarachchi before the main operation began, so that the Tiger cadres will not be able to escape the dragnet the Army was hoping to lay for them. In the grey light of the dawn they dug themselves in, awaiting the progress of the main operation to the north. In the meantime, another batch of commandos had been dropped on the beach around Manalkadu in northeast Vadamarachchi. They occupied the sand banks on the beach to seal that stretch of the coast. Out in the ocean, the navy's gunboats were also on the prowl, ready to provide supporting fire and also to intercept any fleeing rebels.

By this time the entire Jaffna peninsula had been plunged under a dusk to dawn curfew. At first light more Sri Lankan Air Force planes arrived, this time to drop leaflets urging the civilians to take shelter in designated safe havens: government buildings and temples. This was a further

---

[44] 'Killed in Action', *Saturday Review*, 23. 5. 87, p. 1.

[45] Susantha Seneviratne, *Negenahira Mudhagath Meheyuma*, Pahana Publishers, Colombo, 2003, p. 32.

indication of the cautious approach taken by the security forces. Dropping leaflets ahead of an offensive had been first tried during the Short Shrift operations and now it was being implemented again, on a much larger scale in order to minimise casualties to civilians. Then, two hours later, more aircraft swooped on Tiger targets in Vadamarachchi and around Jaffna town. This time they brought deadly cargo with them. "The planes came in at 5 am and they bombed and bombed till 6 pm," one resident later recalled. [46] Another man described how seven helicopters came and shot along the streets. Then three Avro transport planes dropped incendiary bombs while other aircraft strafed and rocketed as they passed. [47]

In Palali where the soldiers waited, the atmosphere was electric. Lieutenant Lucky Dissanayake confided in his wife:

> "The situation here is unbelievable. It is the preparation for war on a scale I have only seen in films. There are about 5,000 troops, armoured cars, armoured vehicles, artillery etc. Yesterday the entire brigade had assembled for review by the General on the adjoining airfield. About 10-15 times bigger than any Independence Day parade, with fighter planes, helicopters, transport planes etc. As the troops in their full battle dress, camouflage, kept marching into the grounds, we in the hospital realized that this was history in the making. This was the first time ever in the history of Sri Lanka that a number of Brigades had assembled. The sight of all of these young boys

---

[46] John Elliott, 'Battle for Tamil Hearts, Minds and Stomachs', *Financial Times*, 6. 6. 87, reproduced in *Lanka Guardian*, 15. 6. 87, p. 8.
[47] Ibid., p. 8.

(18-20) all looking trim, loaded with equipment made us feel proud. The cream of the army was here. The General (GOC of JOC) addressed the officers before the battle. It was reminiscent of the likes of General Patton. He ended his address saying that he assembled here, the best medical teams in the country and that they had nothing to worry. He repeated this at the Commanders conference too. I have gone with him to some of the smaller companies and feel proud to serve with him and his team. Work-wise, so far we have dealt with accidental explosions and misfires but no doubt, we will be stretched soon. Tell the boys (sons) that I am taking part in history and that perhaps one day they too may have the same opportunity...The air is full of expectations, but the morale is high."[48]

Brimming with such confidence the soldiers began their advance around 8.30 in the morning. As they did, heavy mortars and artillery opened up from the camps around Jaffna while naval craft patrolling the shores pounded rebel targets, joining the aircraft already at work. Already desultory firing had broken out at the edge of the advance. The rebel sentries had become alerted to the enemy's stirring to the west and they had begun to exchange fire with the advance elements of the two brigades and the soldiers in the Thondamanaru camp. It had little impact on the advance but it showed that if the operation was meant to be a surprise it was wishful thinking.

The real struggle began after Thondamanaru. The two brigades reached Thondamanaru without loss but here their

---

[48] Quoted in Ranatunge, *Adventurous Journey*, p. 136.

progress hit its first obstacle; the rebels had blown up the Thondamanaru bridge months ago and the brigades had to cross the lagoon using alternative means. Resistance was also beginning to pick up as the engineers began to work on an alternative crossing. Young men armed with AK47s opened up from their bunkers overlooking the advancing columns hoping to pin down the soldiers. But it was not sufficient to deter the advance; the blown up bridge only deterred vehicle traffic but the infantry could wade the shallow lagoon without any great difficulty. And although the army had mustered a fleet of armoured vehicles, the initial advance into Vadamarachchi was an exclusively infantry exercise, companies leapfrogging each other into enemy territory. But this only brought them into the midst of a well-laid minefield. The going became tough here, many soldiers falling victim to the deadly mines liberally strewn in the path of advance. The Tigers had set up the minefield ingeniously, stringing several mines together so that one set off a series of explosions. The leading company under Captain Bahar Morseth took the brunt of the mines, the leading platoon losing nearly ninety percent of its soldiers within the first half hour.[49] One took the lives of eight men from the Engineering Regiment attached to the companies.[50] The command vehicle of the 1st Battalion of the Gajaba Regiment was also blown up; luckily for the commander he was on foot at the time.[51] For a few hours the advance hung in the balance, the troops taking several hours to negotiate the minefield, the final breakthrough coming only in the afternoon.[52]

---

[49] Interview with Brigadier Bahar Morseth, 21. 11. 15..

[50] 'Operation Liberation Encircles Key LTTE Post', *Daily News*, 28. 5. 87, p. 1.

[51] C.A. Chandraprema, *Gota's War: the Crushing of Tamil Tiger Terrorism in Sri Lanka*, Colombo: Ranjan Wijeratne Foundation, 2012, pp. 163-4.

[52] Seneviratne, *Negenahira Mudhagath Meheyuma*, pp. 33-4.

In the meantime the second Brigade had also become active. They had begun making brief forays from the camps in Valikamam including the Jaffna fort. They were also being beefed up by two groups of commandos numbering about 300 landed by sea.[53] To the south, another column – also part of the second brigade - advanced from Elephant Pass. It moved up to Iyakachchi and then on towards Sornampattu with the final goal being Chempionpattu on the Vada strip.[54] They were hoping to seal off the southern end of Vadamarachchi. The column from Elephant Pass overcame initial resistance to achieve its objective. But the troops from the Jaffna Fort ran into difficulties within minutes. They were confronted by Tiger cadres led by their Jaffna commander Kittu and after about two hours of fighting the first column returned to the Fort. A second team advanced about 700 meters and past two street intersections but returned to base after a sharp fire fight. They claimed to have killed ten rebels while losing three of their own. The rebels admitted to losing eight men and claimed to have inflicted heavy casualties on the army. They also claimed to have foiled another foray towards Keerimalai from Kankesanthurai army camp and another attempt by troops to come out of their outpost in the cement corporation at the same location.[55] The Tigers hailed this as a success but the troops had achieved their objective: to keep the enemy guessing.

Meanwhile in southern Vadamarachchi things were not going well for the commandos. As they advanced cautiously with a screen of snipers in front the Tigers detected them and began lobbing mortars, killing lance

---

[53] Iqbal Athas, 'The Vadamarachchi Landing', *Weekend*, 31. 5. 1987. P. 6, O'Ballance, *Cyanide War*, p. 82.

[54] Athas, 'The Vadamarachchi Landing', *Weekend*, 31. 5. 1987. p. 6.

[55] R.C., 'All Quiet on the Northern Front', *Saturday Review*,30.5.87, p. 4, 'Tigers Hold Troops as Captive Force', *Weekend*, 21. 6. 87, p. 11.

Corporal Witharana and Sergeant Dissanayake instantly. They fell back, only three out of the nine men in the group returning unscathed. They settled for the night in Karaveddy and awaited further developments to the north.[56]

To the north however, the advance continued. After Thondamanaru it gathered pace, the Tigers now retreating to their defence lines further to the rear. Still the landmines continued to harry the troops, their threat now compounded by the numerous booby traps the troops had to deal with. The rebels, either in anticipation of a major push in the direction of Vadamarachchi or as a reaction to it, had turned many of the buildings along the path of the advance into veritable death traps. According to one officer's estimate every fifth house had a booby trapped light switch; as soon as the switch was turned on the walls came tumbling down. At times, places where the soldiers could take cover were also booby trapped. At one point of the advance a hail of fire threw the men of the Gemunu Watch in the First Brigade into nearby shelters which immediately erupted in booby trapped explosions injuring several men.[57]

The First Brigade, moving closer to the coast and therefore in close proximity to the main population centres faced the greatest resistance which stiffened as they approached the road that led from Velvetithurai to Uduppiddy. Here, the Tigers rallied under Soosai, their Vadamarachchi leader, fighting back from a line of bunkers built parallel to the road. The Tigers had turned many of the houses in to strong points with bunkers built at ground level as well as underground. Now they fired from these, pinning the soldiers down with small arms, RPGs and 40mm grenade

---

[56] Thilak Senanayake, *Vadamarachchi Vimukthi Meheyuma Saha Uthure Satan*, Colombo: Godage Publishers: 2004, p. 34.
[57] Seneviratne, *Negenahira Mudhagath Meheyuma,* pp. 35-6.

launchers. Again it seemed that the advance would stall, several soldiers being hit, some fatally. But with the help of their heavy weapons the soldiers rallied. A flank attack with RPGs soon convinced the Tigers to abandon their posts but not before they had killed seven soldiers and wounded ten others. The number of rebels killed was not known but their abandoned bunkers showed trails of blood, signs of the dead and wounded being dragged away.[58]

The bunker line breached, the troops performed a new manoeuvre; the First Brigade turned 90 degrees and marched northward, hoping to trap Velvetithurai in a pincer movement. The movement, while it succeeded in taking the town also resulted in tragedy for the Army. Two units of the army at either end of the advancing line mistook each other for the enemy and began firing on each other. Climbing the upper storey of an abandoned house to survey the scene, Captain Shantha Wijesinghe was struck in the chest by a bullet. He was rushed to hospital but died upon admission.[59] The death of Captain Wijesinghe was a huge loss to the Army. Two and a half years ago at Kokilai, it was he who had held out against the first ever rebel attack on an Army camp, earning the reputation as a brave officer. Now he paid for his lack of caution.

---

[58] Ibid., pp. 38-41, Senanayake, *Vadamarachchi Vimukthi Meheyuma*, pp.15-6.

[59] Seneviratne, *Negenahira Mudhagath Meheyuma*, p. 45, Gerry de Silva, *A Most Noble Profession*, p. 78. According to Shantha Dissanayake who was a second lieutenant at the time, Captain Wijesinghe was killed by a solitary mortar round or 40mm round while carried out reconnaissance of Velvetithurai after its capture. Dissanayake claims to have been present at the time. It is inconclusive whether Wijesinghe was felled by friendly fire or enemy fire. I have followed the narrative of De Silva and Seneviratne as they corroborate each other. See Shamindra Ferdinando, 'Army Loses Hero of Kokilai Battle', *The Island*, 10. 7. 13, http://pdfs.island.lk/defence/20130710_153.html

In the meantime, the navy attempted a seaborne landing on the coast of Velvetithurai. But the troops came up against such strong resistance that only one boatload of soldiers could be landed, a sailor being fatally struck by rebel fire in the process.[60] While the soldiers on the beach kept their heads down the gunboats blasted the enemy defences with their cannon and machineguns. Inside Velvetithurai itself sporadic resistance continued, a grenade fired from an improvised grenade launcher killing a soldier from the Engineering regiment and slightly wounding Captain Bahar Morseth.[61] But by now the Tigers were fighting a delaying action as their cadres were slipping out of Velvetithurai. By the 28th the town was in army hands. In the town the troops discovered an elaborate network of bunkers, those facing the sea having such sturdy walls that the 37mm cannon in the gun boats had barely made a scratch on them.

Rebel resistance petered out after the capture of Velvetithurai. Apart from sporadic gunfire and the odd mortar little else bothered the march. The rebels were clearly on the run. At Tikkam, a few kilometres to the east of Velvetithurai, the troops found one of their training camps hastily abandoned, complete with obstacle courses and firing ranges. Close by stood a house that had been converted into one of the workshops that turned out the Tigers' 'baba' mortars.[62] As the workshops went up in flames the troops advanced. By 1st June they were fast approaching Point Pedro, moving cross country in three lines, breaking through fences, cutting barbed wire and scaling garden walls.[63] There was hardly any resistance now.

---

[60] Interview with Lieutenant Sudesh Ranawaka, 31. 1. 17.
[61] Interview with Brigadier Bahar Morseth.
[62] Seneviratne, *Negenahira Mudhagath Meheyuma,* pp. 48-9.
[63] 'Vadamarachchi Operation: The Missing Generation', *Saturday Review*, p. 4.

Further to the south, Denzil Kobbekaduwa's men had a comparatively less eventful advance. They too struggled at Thondamanaru. The troops crossed the lagoon using the undamaged causeway only to run into a barrage of fie from a concrete pill box commanding their line of advance. Open ground lay in front of the pill box which made it difficult to approach it and the sturdy construction defied bullets and even RPGs. The fire from a recoilless rifle finally managed to breach it and an attack from both flanks carried the fortification which was found to be heavily booby trapped. Immediately behind the pill box stood a house belonging to one of the Tamil officers of the Air Force, also liberally fortified with numerous booby traps and the troops had to climb through the roof in order to enter the house and neutralise them.[64] As with the Third Brigade, such encounters invariably delayed the advance. In one widely publicised incident Brigadier Kobbekaduwa narrowly avoided death at a booby trapped house they had surrounded. One of the soldiers rang the bell only to find the building collapsing in a heap. [65]

Such narrow escapes notwithstanding, the advance continued. The column had reached Uduppiddy by mid afternoon on 27th May and Nelliady was occupied soon after. Here the troops faced heavy resistance as the guerrillas used the cover of many coconut groves in the area and a pick-up mounted .50 calibre gun to resist the advance. The army was able to occupy the locality only after a heavy pounding from the air. The resistance neutralised, the Army set up a camp in the Nelliady MahaVidyalyam. On 31st May Puloli, about 10km to the

---

[64] Interview with Colonel Kapila Ratnayake, 13. 12. 15.
[65] 'The Generals Lay a Trap', *Asiaweek*, 14. 6. 87, p. 21.

south of Point Pedro and an important road junction, was taken.[66] The troops now converged on Point Pedro.

As the two columns converged on Point Pedro the troops holed up in the Point Pedro camp broke out to meet them. A few Tigers still left in the town hit back, lobbing mortars as the troops came out. One of them felled Lieutenant Dharshan Fonseka of the 4th SLLI. [67] His loss marred what was appearing to be a triumphant end to the campaign as Pont Pedro was soon secured without further loss. Here, the commanders of the operation posed for a historic photograph under a mango tree to celebrate what appeared to be the conclusion of that phase of the operation.[68]

Within a week, Vadamarachchi had been taken and declared cleared of rebels. The Army had succeeded in capturing one of the three main administrative divisions of the Jaffna Peninsula. But it had come with a heavy price: 33 killed and over 182 wounded, the biggest loss suffered by the Army in a single operation to date.[69]

On 4th June the second phase of the operation began, with a thrust towards Achchuveli. Through this town lay the road to Jaffna less than 20 km away. The first phase of the operation ended with the capture of Iddaikadu. After a brief resistance, the rebels fell back on Achchuveli, blasting culverts as they retreated. At the Achchuveli Maha Vidyalayam they made a stand. The advance came to a halt

---

[66] Senanayake, *Vadamarachchi Vimukthi Meheyuma*, p. 42.

[67] Ibid., p. 24, Ruwan Jayatunge, *Sangramayen Pasu: Ealam Yuddhayata Muhuna Dun Soldaduwange Anuwedaneeya Katha*, Colombo: Sarasavi, 2007, p. 299.

[68] Chandraprema, *Gota's War*, p. 165.

[69] This was the official casualty figure as it appeared in the operation debrief. Cited in . Jagath P. Senaratne, *Sri Lanka Armoured Corps: 60 Years' History*, Sri Lanka Armoured Corps, 2015, p.117. O' Ballance, *Cyanide War*, p. 84 gives the figure of 37 killed and 168 wounded.

here, Captain Navaratne of one of the attacking companies being hit by a rifle grenade. As Navaratne breathed his last, mortars from the Thondamanaru camp opened up, squashing the resistance. The advancing soldiers found the bodies of a dozen rebels lying about the school. Some had taken cyanide and two of them were still breathing.[70]

With the capture of Achchuveli, Operation Liberation ended. The troops withdrew from the locality a few days later. At the time Brigadier Gerry De Silva who headed the operation gave the reason for the withdrawal as the concern for civilian casualties.[71] But the reality was that by this time there were also other forces at work, dooming the operation. On 4th June soldiers returning to base from the Vadamarachchi operation had observed two large transport planes escorted by four Mirage fighters circling the sky over Jaffna. Even though they were within firing distance the soldiers were ordered to desist from engaging the planes.[72] The aircrafts were on a mission from India. Despite their annoyance with the intransigence of the Tigers, the Indian government had not been happy with the launching of the offensive which it saw as an attempt to crush the militants for good and thereby undermine the influence India had over Sri Lanka. And as news of widespread destruction in Vadamarachchi and civilian casualties began to pour in, India issued vehement protests. When they failed to stem the military advance the Indian government sent a flotilla of fuel, food and medicines to Jaffna as a "humanitarian gesture." The Sri Lankan navy turned the flotilla back on June 3 but the following day India retaliated by sending five Antonov transport planes

---

[70] Seneviratne, *Negenahira Mudhagath Meheyuma*, pp. 60-3.
[71] Derek Brown, 'Jaffna Reality – Two Strange Forms', *The Guardian*, 15. 6. 87, reproduced in the *Saturday Review*, 11. 7. 87., p. 4.
[72] Seneviratne, *Negenahira Mudhagath Meheyuma*, p. 59.

with relief supplies escorted by four Mirage fighters. The planes entered Sri Lankan airspace for a few minutes, dropping 22 tonnes of relief supplies. The message to the Sri Lankan government was clear. The military offensive would continue only at the risk of inviting further Indian interventions.

Thus, Operation Liberation came to a premature end with the troops in control of Vadamarachchi but unable to move beyond. The order to halt the operations came as a great disappointment to the men who had taken part in the drive anticipating a decisive campaign. They had to be content with holding Vadamarachchi awaiting further diplomatic and political developments. In the South there was outrage. When the Indian premier Rajiv Gandhi arrived in Colombo to sign the Indo-Lanka Accord which followed the Indian intervention there were bloody riots in the capital. It was hardly the result the government had expected.

CHAPTER 6

Victory Denied? 'Operation Liberation' in Retrospect

Operation Liberation was a watershed for the Sri Lankan security forces. For the first time in their short history the Sri Lankan armed forces had launched a combined arms operation involving several brigades. It required sound planning and good co-ordination and tested the ability of the armed forces to conduct offensive manoeuvres over a large area, some of which was built up environment. The challenge was not only to trap and destroy the enemy and his resources but also to do so without causing unnecessary harm to the civilians. Despite all the bravado of the government, a large loss of civilian life was to be avoided. And all this had to be achieved with technological means that were backward, even by the standards of the developing world.

6.1 The Balance Sheet

In terms of capturing territory and crushing enemy resistance the operation proved a success. The rebels were no match for the Sri Lankan security forces if they made a determined and concerted effort. Despite all their much vaunted ability to confine the Army to the camps the Tigers failed to stop the advance. They were facing an assault from several directions for the first time and they found that they wilted under the pressure, the firepower of their heavy machine guns and home-made mortars not sufficient to halt brigade-strength thrusts bolstered by heavy artillery, armour and air cover, no matter how primitive some of these arms may have been. At Thondamanaru when it was evident that the army had broken through the mine barrier the Tigers fell back telling the civilians to run for cover. A resident of Kamparmalai put the disparity into perspective. "I will not say that the boys didn't fight hard" he told the *Saturday Review*. "The firepower of the army was immense. You cannot stand up to shells with AK47s."[1]

However, the area actually 'captured' was small. The government declared all of Vadamarachchi cleared of rebels but the area over which the security forces enjoyed near total control was a small oblong area in the north of the Vadamarachchi sector bordered by Velvetithurai, Point Pedro, Uduppiddy and Nelliady. This was indeed the most populous part of Vadamarachchi but much territory lay outside it, territory in which the rebels were relatively free to roam. They could also move back and forth between Vadamarachchi and other sectors. The coast was also not completely under control. The long coastline southward from Point Pedro was largely open.

And the Tigers seemed to have made good use of these opportunities. The security forces claimed to have killed 96 rebels and wounded 130 but independent confirmation of this was not available.[2] The Tigers put their own losses at a paltry 13 killed with about 20 missing from EROS.[3] A large number of young men were rounded up and sent into

---

[1] 'Vadamarachchi Operation: The Missing Generation', *Saturday Review*, 20. 6. 87, p. 3.
[2] . Jagath P. Senaratne, *Sri Lanka Armoured Corps: 60 Years' History*, Sri Lanka Armoured Corps, 2015, p.117. Edgar O'Ballance, *Cyanide War: the Tamil Insurrection in Sri Lanka 1973-88*, London: Brassey's, 1989, p. 200 gives rebel casualties according to army sources as more than 200 killed.
[3] The Day of the Lions (Jackals?)', *Saturday Review*, 6. 6. 87, p. 2.

captivity awaiting interrogation. But it is likely that many of the hardcore of the rebel cadres escaped death and capture. As the troops consolidated their gains at Velvetithurai word came from the commandos that many of the Tiger cadres had filtered out of Vadamarachchi using little known paths.[4] Later an army officer confided in a journalist that not only the Tiger chief but also 300 – 350 top cadres of the LTTE had made their escape largely due to the inability to seal off the coast completely, especially the long stretch of coast in eastern Vadamarachchi.[5] Those that escaped included Prabhakaran and Soosai. It was later revealed that the Tiger supremo had slipped out of the net when the troops approaching the area he was holed out in were delayed by booby traps.[6] He is said to have escaped through the Vallipuram area which was to be blocked by the commandos.[7] Moreover, the army's haul of weapons was also small and included few heavy weapons. A mortar factory with a large haul of bombs was found in and a few mortars were also recovered but these were mostly home-made items and could be easily replaced. The pick-up mounting the .50 calibre machine gun was recovered but without the weapon suggesting that it was either hidden or removed to safety.[8] The capture of territory and the smashing of resistance did not accompany the destruction of the enemy's forces or his arsenal.

---

[4] Susantha Seneviratne, *Negenahira Mudhagath Meheyuma*, Pahana Publishers, Colombo, 2003, p. 48.
[5] Qadri Ismail, 'Military Option and its Aftermath', *Sunday Times*, 7. 6. 87, p. 5.
[6] Rajan Hoole et al., *Broken Palmyrah, The Tamil Crisis in Sri Lanka, an Inside Account*, Sri Lanka Studies Institute: Ratmalana, Colombo, 1992, p. 125.
[7] Gerry de Silva, *A Most Noble Profession: Memories that Linger*, Colombo: International Book House, 2011, p. 80.
[8] Interview with Colonel Kapila Ratnayake, 13, 12, 15..

These shortcomings in the operation were hard to avoid. The armed forces were still getting their act together and taking part in their first major combined-arms offensive. The commandos were still small in number and learning their trade. In later years they would develop into a potent force, capable of handling rebel cadres far more deadly than those they were tasked with containing in Vadamarachchi and in terrain far more challenging. But at the time they were still in their infancy, struggling to live up to their expectations. Likewise the Navy too had very limited resources with which to patrol the long coastline and had to be content mainly with covering the northern stretch of the Vadamarachchi coast.

In fact, in hindsight, the conception of the whole operation appears amateurishly ambitious. Two columns of troops moving West - East were expected to tie down an enemy that was relying mainly on landmines and booby traps while less than a hundred commandos and Special Forces troops were tasked with keeping a large area under surveillance. It was a sign of a military command grappling, for the first time, with the challenge of overseeing multi-brigade campaigns with scarce resources and ambitious expectations.

The advance itself suffered at times from the inexperience of the troops. They struggled to maintain alignment causing gaps between them and making the flanks of the units vulnerable to attack and infiltration. Sometimes, in the confusion of the advance, troops were prone to fire on each other mistaking their own troops for the enemy. The death of Shantha Wijesinghe seems to have occurred due to such a mistake. Such mishaps occur in more experienced armies too but it is likely that a more professional army would have been able to handle an operation of this scale without too much difficulty.

The air force provided valuable cover to the advancing troops but at times the reluctance of pilots to get too close to the enemy fire was evident. The concerns of the air force were betrayed when air chiefs refused to fly after ground fire became too threatening. President .Jayawardena revealed in 1990 that the security chiefs had informed the cabinet that an aircraft had been lost to rebel fire and if one more aircraft is lost the Air Force would refuse to fly any further missions.[9] Even during the operation the Air Force had often taken care to stay clear of rebel fire sometimes testing the patience of the ground troops. Colonel Vijaya Wimalaratne is said to have remarked in despair that he had no power to command a helicopter to land to evacuate casualties. The casualties had to be removed to Thondamanaru where they were picked up by the helicopters.[10]

However, an even bigger concern was the amount of suffering unleashed on the civilians by the scale of the operation and the deployment of heavy firepower. The number of civilians killed is hard to establish. Government sources gave the civilian casualties at 47 while the Tigers claimed hundreds were killed. The government no doubt, underplayed the extent of the harm to civilian life and property while it was in the Tigers interest to exaggerate them. Indian sources, often fed by pro-Tiger informants and with India's own interest in bringing the hostilities to an end in mind, also joined the chorus accusing the military of causing heavy civilian casualties. There was also the difficulty of differentiating between civilian and rebel casualties

---

[9] "J. R. Breaks His Silence: I Feared a Military Coup in'87." Interview with Vijitha Yapa, Sunday Times, 11. .2. 90, pp. 14-15. Jayawardena says the threat was made after an air craft had been shot down but to the best of my knowledge this is not confirmed. As generally accepted the first air craft lost in the war was Siai Marchetti that crashed into the Jaffna lagoon during the operation to break the Tigers' siege on Jaffna in August-September 1990.

[10] Tilak Senanayake, *Vadamarachchi Vimukthi Meheyuma Saha Uthure Satan*, Colombo: Godage Publishers, 2004, p. 16 and p. 18.

as many of the rebels wore civilian dress and given the army's well-known propensity to pass off civilian dead as 'terrorists.' But it is clear that there was widespread destruction from the bombing and shelling. A Western journalist, one of the few independent observers to visit Jaffna at the time, found no evidence of "carpet bombing" as claimed by some Tamil sources but witnessed "line after line of buildings beyond repair" in Velvetithurai which had borne the brunt of the government's assault.[11] Point Pedro was in no better shape. "Between the army camp and the inhabited area there lies a swathe of utter destruction" observed another journalist. "The main square is littered with rubble and whole buildings have collapsed."[12]

The army attributed much of the destruction to the booby traps left by the retreating Tigers. There is no doubt these caused much damage, to property as well as to the advancing troops. It is also not certain if all the buildings destroyed during the offensives were occupied at the time. An Indian visitor to Jaffna in February 1987 found many of the houses in Velvetithurai abandoned by their residents.[13] Many of the remaining residents would have also taken refuge in the designated places. However, some reports spoke of civilian casualties caused by aerial bombing and shelling. According to a foreign journalist heavy mortars fired from the Jaffna Fort and naval gunfire caused civilian deaths and spread terror without warning. "Within hours (of the leaflet drop) the downtown area was hit by heavy mortar fire and shells from a naval gunboat. The result was panic and confusion among the civilian population – one mortar that landed near the hospital killed a 12-year-old

---

[11] John Elliott, 'Battle for Tamil Hearts, Minds and Stomachs', Financial Times, 6. 6. 87, reproduced in *Lanka Guardian*, 15.6.87, p. 8.
[12] Derek Brown, 'Jaffna Reality – Two Strange Forms', *The Guardian*, 15. 6. 87, reproduced in the *Saturday Review*, 11. 7. 87, p. 4.
[13] S. Venkatnarayan, 'A Visit to Jaffna', *The Island*, 15. 2. 87, p. 6.

boy. A single small piece of shrapnel had hit him in the right side of his head."[14] Another Western journalist found the Puloli hospital full of wounded civilians. A doctor told him that 20 civilians had died in the first four days from wounds received.[15]

According to other sources, the places of worship which were designated places of refuge did not escape the shelling and bombing either. One shell landing on Mariamman temple in Alvai is said to have killed 35 people and another bomb hitting the Sivan temple on Kankesanthurai Road claimed 17 victims.[16] Eight people including women and children were reported killed by a shell at St. James School while 7 more were reportedly killed by a bomb at the Amman Kovil at Suthumalai.[17] Reports claimed up to 75 people being killed by shellfire were burnt to death at Mathumari Amman temple at Alvai on 29th May[18].

These excesses are not surprising given the primitive resources at the disposal of the Sri Lankan military. The armed forces had improved immensely in terms of equipment and training since 1983 but they still lacked the sophistication to deal effectively with a largely urban guerrilla movement. The heavy firepower of the armoured cars, heavy artillery and RCLs were potent weapons against structures and enemy concentrations but given that the Tigers had few of these they could only pulverise

---

[14] Robert McDonald, 'Eye Witness in Jaffna', *Pacific Defense Reporter*, August 1987, p. 27.

[15] Jon Swain, 'Children Suffer Horrific Burns in Army's Offensive Against Tamil Guerrillas: Sri Lanka Ends Fighting to Give Peace a Chance', *Sunday Times*, 14. 6. 87. The doctor also confided the lack of drugs and the emigration of the surgeon were main contributors to the deaths but the wounds of the deceased seem to have been caused by the bombing.

[16] Hoole et al., *Broken Palmyrah*, pp. 127-8.

[17] 'Operation Blue Star', *Saturday Review*, 30. 5. 87, p. 1.

[18] 'The Day of the Lions (Jackals?), 'Vadamarachchi Operation, The Missing Generation', *Saturday Review*, 20. 6. 87, p. 3.

indiscriminately. The artillery in particular lacked accuracy, not being able to do more than laying down fire over a large area in the general direction of a target. In the air a similar story prevailed. The Air Force had mustered an impressive array of aircraft but as we have seen, their potential for engaging the enemy effectively was very limited. Once again the Tigers provided very few identifiable targets and even if such targets could be identified, the helicopter gunships, ground attack aircraft and the improvised 'bombers' had no means of engaging them with any precision. As a consequence, civilian casualties were to be expected, if not inevitable. As explained earlier, the helicopters were cumbersome machines that were hard to manoeuvre and the ammunition was often unreliable as to accuracy. The low quality ammunition had an equal chance of hitting the target as well as bystanders, especially in built up areas where the militants were often encountered. Sometimes helicopter crews resorted to a crude form of bombing, dropping hand grenades stuck in wine glasses after their pins were released.[19] When the glass broke the lever was sprung setting off the explosion. The rockets fired and bombs dropped by the Siai Marchettis were no more accurate.

However, the most inaccurate and terrifying weapon was the barrel bomb, a new and savage development in an ugly war and evidently unveiled for this campaign. These were large metal 45 gallon drums filled with gelignite or sometimes flammable gas or rubber tubes. Sometimes they were also filled with explosives. Upon exploding, the flaming pieces of rubber burst out sticking to the skins of anybody unfortunate to be in the vicinity.[20] The barrel bombs were even more inaccurate than the rockets and the machine guns. They were dropped (or pushed – sometimes

---

[19] Michael Hamlyn, 'Slow Struggle on Jaffna Peninsula', *The Times*, 1. 6. 87.
[20] Thomas Abraham, 'Pounding Jaffna', *Frontline*, 2-15.3.91, p. 56.

using feet) from a height of over 300 metres to stay out of range of the Tiger .50 calibre machine guns and having no ballistic stability whatsoever they could hit anything within a wide radius of the target.[21] The Sivan temple was reportedly hit by one of them killing a number of civilians and destroying the historic building while many of the wounded in the Puloli hospital observed by the aforementioned Western journalist had suffered horrific burs from a barrel bomb.[22] It was a danger not only to the foe but to friend as well. "If you look up you can see them twisting and turning as they fall," said one army colonel to a journalist. "Sometimes we ourselves are mortally afraid of where they're going to land."[23] The terrifying inaccuracy of the bombs can be gauged from the fact that in late 1990 during the LTTE's siege of the Jaffna Fort, some of the barrel bombs dropped by the air force on targets around the fort fell within the walls of the fort.[24]

Given these primitive means of carrying out air raids it would have been a miracle if civilians had not been hurt. However, what is disturbing is that many bombs and rockets seem to have fallen on the very places that were designated as safe leading to speculation that the attacks were deliberate. The Tigers' propensity to take position near civilian centres cannot be discounted either. Throughout the siege of the Jaffna Fort they had not hesitated to attack the camp from locations surrounded

---

[21] McDonald, 'Eyewitness in Jaffna', p. 26.

[22] Jon Swain, 'Children Suffer Horrific Burns in Army's Offensive Against Tamil Guerrillas: Sri Lanka Ends Fighting to Give Peace a Chance', *Sunday Times*, 14. 6. 87.

[23] Julian West, 'Passage to Jaffna', *Asiaweek*, 8. 3. 1991, pp. 17-18. Even though West was writing in 1991 during Ealam War II, the above observations are valid for the first phase of the war as well, considering that the same technology was being used.

[24] Gamini Goonetilleke, *In the Line of Duty; Life and Times of a Surgeon in War and Peace*, Colombo: Unigraphics, 2008, pp. 113-14.

by civilians. It would have come as no surprise if they had done so during the offensive. At the same time the Air Force also has had a record of indiscriminate and even wilful targeting of civilians. All this formed a deadly combination which only served to terrorise and harm the civilians.

The security forces did, however, develop a primitive yet ingenious way of obtaining a degree of accuracy with bombing and shelling. Suspected rebel targets were marked with circles on a map with each circle given a number. The officers on the ground, the air force and the artillery operators shared the maps (known as 'bola' maps). When bombing or shelling of a particular target was called for, the number of the corresponding circle was relayed to the air force or the artillery and the location was bombed or shelled.[25] This was no doubt a genuine effort to minimise civilian casualties and ensure greater accuracy in the absence of more sophisticated means but it was not sufficient to obviate the inherent drawbacks of the weapons deployed.

Apart from the limitations in the weapons and the delivery systems, the unprofessional conduct of some troops also endangered civilian life. There were reports of soldiers firing at random into houses and bunkers sheltering civilians as they advanced past them.[26] Such actions seem to have followed a simple, if ruthless, rationale. People who remained behind when they had received clear instructions to move to safe places were either 'terrorists' or "terrorist sympathisers" and therefore, fair game. What was not grasped was that many people found the safe havens too far from their homes and the temples and the government buildings were not adequate to house everybody. Besides, there was also the simple terror that made many people remain at home. But the soldiers moving cross country had little time or desire to consider their point of

---

[25] Interview with Colonel Kapila Ratnayake, 13. 12. 15.
[26] 'The Vadamarachchi Operation: the Missing Generation', p. 4.

view. They were also nervous as suggested by the random firing into houses. Remarkably, some civilians empathised with them: "A lot of people were killed, but I think they were foolish," said one man. "They were sheltering in their bunkers and when the shooting started they looked out to see what was happening. In the dark a soldier can't see who is a terrorist and who is not, so naturally they shot them. "[27]

However, there is also evidence of a more sinister type of violence against civilians in the form of the deliberate targeting of non-combatants, often in order to terrorise. While the advance continued, men in sarongs appeared in some areas of Vadamarachchi occupied by the security forces, carrying swords and knives. Many civilians were reportedly hacked to death by them. According to one report as many as 80 people were killed in this fashion around Nelliady.[28] There were also damning reports of summary executions and indiscriminate shooting of civilians in cold blood. On one such occasion a group of civilians who were being led away in handcuffs were asked to run and then mowed down.[29]

Such actions pointed to a continuing amateurism in the Sri Lankan Army. Again, this is not surprising. Despite having fought the rebels for half a decade they were still strangers to large scale operations especially in areas heavily frequented by civilians. Coupled with the enthusiasm for teaching the enemy a lesson and their own nervousness at being shot at, such inexperience provided a good base for callousness towards civilians who got in their way or who were suspected of rebel affiliations.

---

[27] Ibid.
[28] Ibid., 'The Day of the Lions (Jackals?)'.
[29] Hoole et al., *Broken Palmyrah*, p. 128.

6.2 The Guerrilla Challenge

Even in combating enemy resistance, the security forces were not completely unchallenged. The army had crushed rebel resistance with heavy firepower but this was when the Tigers had decided to stand and fight. When they resorted to what was their real strength, guerrilla warfare, the armed forces found themselves struggling. The vast majority of their casualties were not caused by small arms or mortars but by booby traps and mines. But the prowess of the rebel as a guerrilla became clearer in the days following the capture of Vadamarachchi. Having failed to stop the army's advance the Tigers hit back at the enemy with guerrilla attacks on their lines of communication and outposts. Soon after Operation Liberation ended a landmine blew up a truck bringing released detainees back to their homes, killing ten of the detainees and three soldiers.[30] Around the same time an officer confided in a foreign journalist that "some terrorists" had infiltrated the area considered cleared by the Army.[31] However, the first major counter blow came in Jaffna town. In the small hours of 3rd June, an improvised armoured truck laden with explosives approached the military post at the telecommunications building adjoining the Jaffna fort. The truck was the spearhead of a devastating attack by a group of Tigers said to have numbered more than fifty. The explosion brought down much of the building killing three soldiers and wounding more than forty. The Tigers attacked the survivors furiously and the soldiers fought back with the aid of helicopters that sprayed the area with machine gun fire and rockets. By morning the attackers had withdrawn, taking their dead and wounded with them. Among the booty captured by the rebels was one GPMG, seven Belgian made FNGs, one AK grenade launcher and over

---

[30] Iqbal Athas, 'The Post Vadamarachchi Crisis', *Weekend,* 14.6.87, p.6.
[31] Derek Brown, 'Jaffna Reality – Two Strange Forms'.

a 1000 rounds of ammunition.[32] Tigers claimed to have killed 22 soldiers, including a second lieutenant.[33] More importantly, they also took captive three soldiers, their second such haul that year. On 6th June, the *Saturday Review* published a picture of the three young men, barely out of their teens[34].

The biggest counter strike came at Nelliady on 5th July. The army had captured Nelliady during the first few days of Operation Liberation and had set up a camp in the Nelliady Central College. About 100 men from the Gemunu Watch garrisoned the camp supported by a troop of armour (a Saladin armoured car and two scout cars).[35] The troops were relaxed, considering the town to be cleared of militants. That night however, the Tigers proved them wrong. Shortly after 8 pm the soldiers in the school found mortars and rockets raining on them. The Tigers had slipped into the row of abandoned buildings opposite the school and were using their cover to stage an attack. But the main attack was approaching down the road. Having cleared the road blocks along the approach road leading to the school with rockets and rifle grenades, the Tigers drove down two explosive laden trucks into the school. One overturned and exploded prematurely but the other, driven by a rebel cadre named Miller rammed into the school and burst into flames bringing down a part of the school. Making use of the chaos and the breakthrough made by the explosion, several militants dashed in to the compound but the soldiers fought back, resisting fiercely. As the troops

---

[32] Iqbal Athas, 'How Operation Jellyfish Stung India's Flotilla at the Kutch', *Weekend*, 7. 6. 87, p. 23.

[33] 'Telecom Soldiers Charred', *Saturday Review*, 6. 6. 87, p. 8.

[34] 'Cannon Fodder', *Saturday Review*, 6. 6. 87, p. 1.

[35] Shamindra Ferdinando, 'Tragedy in the Wake of Victory at Velvetithurai', *The Island*, 8. 1. 13,
http://www.island.lk/index.php?page_cat=article-details&page=article-details&code_title=69950

engaged the militants a sergeant and a driver got into the Saladin armoured car in the schoolyard. A few rounds from its 76 mm gun would have flattened the buildings and relieved the pressure on the camp. But chance had it that an RPG found its way into the armoured car through the driver's viewing aperture incinerating him and the sergeant and putting the vehicle out of action.

The fire-fight continued for several hours. Reinforcements sent in helicopters struggled to land their troops amid heavy gunfire while troops approaching by road had to negotiate freshly laid mine fields. Finally the troops battled their way to the camp and the Tigers seem to withdraw. However, while the crumbled masonry was being cleared in search of bodies the attack recommenced, a mortar hitting the cook house and small arms fire erupting again. The firing eventually died down leaving the exhausted soldiers to count the dead and wounded. Officially, military's death toll stood at 17 with over 30 injured. But it was probably much higher. The Tigers admitted to losing three men including the driver of the truck, "Captain Miller", the first suicide 'martyr' for the rebels. They also captured a large haul of small arms and ammunition including a mortar launcher.

That however, was not the end of Nelliady's trials. The following Thursday (9th) Tigers fired mortars at Nelliady again, wounding ten soldiers. The day after the army's forward base at Polikandy also came under attack, costing the life of one soldier and wounding another[36].

A few days later on the 11th July the army launched an operation to flush out infiltrated militants from the Vadamarachchi area. Dubbed "Operation clean Sweep" it cost the army three dead and twenty three wounded while claiming

---

[36] For the Nelliady attack see Iqbal Athas, 'Tigers Explode Peace a Nelliady', *Weekend*, 12. 7. 87, p. 6, D. B. S. Jeyaraj, 'Birth and Growth of the Black Tiger Suicide Squad', The Island, 13. 7. 97, p. 16,

the lives of eighteen Tigers.[37] Not long after the Kurumbasiddy mini camp near Palali airbase came under mortar attack demonstrating the resilience of the rebel guerrillas.[38]

The operation also exposed another problem the armed forces faced in combating the rebels. The capture of Vadamarachchi required a major commitment of resources which left the East more vulnerable to rebel attacks. The Tigers demonstrated this in telling fashion on 2nd June. While the troops were consolidating their positions in Vadamarachchi, they struck at Aranthalawa, waylaying a bus transporting thirty two young Buddhist monks and slaughtering them[39]. It demonstrated the complexity of the challenge facing the security forces that had to use their limited resources to cover a vast theatre of operations against an elusive enemy.

6.3   Victory Interrupted?

The Vadamarachchi campaign showed the challenges the inexperienced Sri Lankan security forces faced in trying to capture and hold territory from a predominantly guerrilla enemy using a primitive conventional arsenal. It is not certain if the campaign was the first phase of a grand campaign to capture the whole peninsula. It has often been so interpreted, especially in the aftermath of Indian intervention which has come to be seen as undermining the triumphant advance of the

---

[37] Iqbal Athas, "The 'Day After' Mood Plagues Jaffna', *Weekend*, 19. 7. 87, p. 6.

[38] Iqbal Athas, 'Island Pride and Prejudice', Weekend, 24. 7. 87, p. 6.

[39] Austin Fernando, 'Revisiting Aranthalawa Bhikku Massacre', *Colombo Telegraph*, 2. 6. 12, https://www.colombotelegraph.com/index.php/revisiting-aranthalawa-bhikku-massacre/

army.[40] However, given the challenges faced by the Security Forces in capturing and holding Vadamarachchi it is questionable whether they would have been able to repeat their success in the rest of the peninsula, especially in Valikamam even if India had turned a blind eye to the offensive. The key challenges faced by the security forces in taking Vadamarachchi were likely to recur on a larger scale and with greater intensity in Valikamam. The Tigers had the hardcore of their forces guarding Jaffna town where they also had the bulk of their heavy weapons. One Western observer in Jaffna at the time noticed an influx of brand new G3 rifles and .30 calibre machine guns along with at least two new .50 machine guns and surmised that these were provided by India.[41] This was a possibility; or it may be an overzealous interpretation by the observer. What is more clearly established is the arrival of a large consignment of explosives – about 20 tonnes - from the Indian Intelligence agency RAW in June.[42]These supplies would have boosted the rebels' arsenal; the explosives would have played their part in the Nelliady attack. But even without such contributions from India the Tigers had sufficient firepower and manpower in Valikamam to offer stiffer resistance to the army's advance than in Vadamarachchi. The fact that Valikamam was more heavily built up than Vadamarachchi afforded the rebels an added advantage. One can only imagine the nightmare of dealing with booby traps in the thickly built up suburbs between Jaffna town and Palali.

---

[40] Gerry de Silva claims that the capture of Achchuveli was the beginning of an advance on Jaffna town with the Palali - Jaffna Road as the axis of advance. He also seems confident that Jaffna would have fallen tot he advancing troops had it not been for the Indian intervention. De Silva, *A Most Noble Profession*, p.80.

[41]The observer, Robert McDonald suspected that these were provided by India. Robert McDonald, 'Eyewitness in Jaffna', p. 29.

[42] Manoj Joshi, 'A Base for all Seasons: how LTTE used Tamil Nadu', *Frontline*, 3-16. 8. 91, p. 22.

That portended heavy casualties for the troops. In this respect, it is instructive to examine the challenges faced by the IPKF just over five months later in capturing the same area. The Indians lost 214 killed and over 700 wounded in capturing Valikamam[43]. To be sure, the Indians laboured under several limitations. They were not allowed to use heavy weapons and the use air cover was minimal. They were also less familiar with the terrain and the units that took on the Tigers were under strength. Their men were also less battle hardened than the Sri Lankan soldiers. Still, all this does not detract from the ferocity and ingenuity of the LTTE's resistance which held up the Indians' advance for more than a week. Given their performance in Vadamarachchi against what was very much a second-string rebel force it is doubtful whether the Sri Lankan forces would have fared any better with all their experience and weaker inhibitions regarding the use of heavy weapons.

And even if they had been able to smash their way into the heart of Jaffna, there was still no guarantee that they would have been able to destroy the enemy forces. There was little chance of preventing Tiger cadres from slipping into the Vanni to fight another day as happened during the overrunning of Vadamarachchi. It is instructive to remember that the Indians also failed to capture any of the rebel leaders when they assaulted Jaffna in October 1987. Equally importantly, an incursion into other sectors of the peninsula also promised severe losses to the civilians. In the heavily built up parts of Valikamam and Thenmarachchi the shells, rockets and bombs of the army and air force could cause more havoc than they did in Vadamarachchi. That could only mean one thing. Massive civilian casualties.

---

[43] This was the official figure after the 16-day campaign to eject the Tigers from Valikamam. Dilip Bob, 'A Bloodied Accord', *India Today*, 15. 11. 87, http://indiatoday.intoday.in/story/after-16-days-of-bloody-battle-ipkf-finally-captures-ltte-stronghold-jaffna/1/337703.html

The army would have been prepared to absorb heavy casualties if it was in exchange for mastery over the peninsula. If there was no threat of external intervention, which would have been the case if India had remained aloof, the government could have also ignored the civilian suffering. However, the destruction of civilian life and property posed another potential threat: the alienation of the population that could only serve to swell the ranks of the rebels. This threatened to aggravate another problem the army had to take into account – the overstretching of their resources.

At least 5,000 troops have been required to take Vadamarachchi and divert the Tigers' attention elsewhere. Many more would have been required to take Valikamam and Thenmarachchi while holding Vadamarachchi. The army had in excess of 40,000 troops at the time but these numbers would have been fully stretched in the event of the occupation of the entire peninsula in the face of a population outraged by the destruction caused by the campaign and therefore, even more supportive of the rebels than they had been before the operation. Such a situation would have been ideal for a protracted guerrilla war, a taste of which the Army had begun to get in Vadamarachchi. When one also factors in the need to keep the vulnerable East under control, the prospects become bleaker. Even if the military success in Vadamarachchi had boosted recruitment pursuing the war on several fronts would have been a daunting task. Athulathmudali himself later admitted that it was questionable whether the government had adequate forces to dominate Jaffna.[44] One can never be certain but it is likely that it was a different kind of war rather than a decisive victory that beckoned the Sri Lankan security forces had the offensive continued after the capture of Vadamarachchi.

---

[44] Rohan Gunaratna, *Indian Intervention in Sri Lanka: the Role of India's Intelligence Agencies*, South Asian Network on Conflict Research: Colombo, 1993, p. 178.

AFTERWORD

The abrupt end to Operation Liberation marked the culmination of the first phase of the military conflict between Tamil insurgents and the Sri Lankan state. This phase, now commonly referred to as Ealam War I, began with individual assassinations and then escalated into attacks on outposts and patrols that led to the establishment of rebel control over much of the Jaffna peninsula. It ended with the Sri Lankan armed forces' attempt to regain control of the Vadamarachchi sector of the peninsula.

The pressure from the insurgency contributed immensely to the evolution of the Sri Lankan armed forces from a largely ceremonial establishment into a fighting force. As the insurgency intensified, the security forces expanded in numbers and acquired new weapons and equipment. They also learnt the rudiments of counter insurgency warfare and urban warfare. The invasion of Vadamarachchi showed that they had also acquired the capability to launch complex, multi-brigade operations where all three arms of the security forces combined.

It had been a learning experience filled with trials and travails. The outbreak of the insurgency had caught the tiny military off guard; the security forces had neither the expertise nor the experience to handle the challenge posed by the Tamil militants. The government's lack of interest in dealing with the underlying issues of the insurgency retarded the development of an effective counter-insurgency strategy. The soldiers' frustration often burst out in retaliation against the civilian population complicating matters. Faced by a hostile population and a burgeoning

185

insurgency that thrived on it, the army found itself increasingly confined to its bases in the north, forfeiting territory to the rebels. What emerged into the plains of Vadamarachchi in May 1987 was, by and large, the result of the struggle to negotiate these challenges and pitfalls.

True, the armed forces were still far from a fully fledged modern professional body. Compared to the situation at the outbreak of hostilities, the soldiers were well equipped and trained in 1987, but the armour and artillery was woefully inadequate. The air force and the navy were primitive by even the standards of the developing world. As shown by Operation Liberation, trying to execute complex, multi-purpose operations in brigade strength, though in itself an achievement, was still a challenge to the armed forces. The army succeeded in overrunning territory but failed to trap and capture or destroy the enemy forces. The security forces were now more mindful of the need to safeguard civilians but civilian casualties still occurred, either due to the primitive weaponry deployed, inexperience of the troops or calculated or casual brutality. But the first tentative, stumbling steps towards becoming a fighting force had been taken.

The scope of this evolution of the armed forces was determined by the nature of the insurgency and the financial and diplomatic constraints placed on the Sri Lankan government. The Tamil insurgency was still at the early stage of the terrorist-guerrilla phase with the insurgents armed with an assortment of small arms, landmines and a few heavy weapons such as heavy machineguns and home-made mortars. Of these, the landmine was the most lethal weapon and this was reflected in the army's acquisition of armoured vehicles better suited to withstanding landmine blasts. The curtailment of road transport largely as the result of mine attacks led to the abandonment of the Jaffna Peninsula to the rebels which in turn necessitated a build up to retake it. The landmine threat and the government's possession of a minute air force encouraged the development of this arm for transport and surveillance as well as retaliation. The clandestine nature of rebel activity necessitated the raising of Special Forces. An area in which the rebels excelled was their naval lifeline to the subcontinent and to combat this, the navy had to be modernised. All this had to be done in the face of great financial hardship and diplomatic obstacles. Money was in short supply and few countries were willing to sell what Sri Lanka needed even when money was available. The result was a poor man's army, navy and air force that were better than what was there before but still below the expectations of the government.

In many ways this development foreshadowed the future expansion and modernisation of the security forces. As explained in the beginning of this book, the present day Sri Lankan armed forces are entirely a creation of the military conflict with the Tamil Tigers, the 1980s marking only the beginning of this journey. The evolution continued after 1987 under pressure from the enemy and aided by changing international opinion. In the 1990s, with the LTTE growing in numbers and firepower, the war turned into a semi-conventional contest with the Sri Lankan forces rising to meet the challenge. The post Cold War world order relaxed the constraints on the Sri Lankan government for obtaining weapons and expertise. The end of the Cold War and the assassination of Rajiv Gandhi reshaped India's attitude to the conflict in Sri Lanka with Delhi increasingly wary and weary of the Tigers. In these changed circumstances it was mainly the financial constraints that inhibited Sri Lankan security forces from modernising as they wished. In the 2000s, particularly after 2005 under increasing pressure from the LTTE and determined political and military leadership Sri Lankan security forces reached the zenith of their power, with the blessings and active support from a wide range of foreign powers, notably India, China and the United States. This power helped the Sri Lankan armed forces defeat the Tigers totally in a relentless offensive which dwarfed the fighting in the 1980s in scale, intensity and destruction. Along with the complexity and intensity of the war the professionalism of the security forces also increased but not sufficiently to dispel allegations of war crimes during the crucial last phase of the war.

In 1987 all this lay in the future. But the groundwork had been done. Pressed by a determined enemy, a largely hostile international community and a parlous economy, the armed forces had learnt to fight, no matter how imperfectly - or savagely. The education will continue as the carnage intensified in the next two decades. But the initiation was over.

List of Images

1. Cordon and search Jaffna, December 1984. In the foreground is a Saladin armoured car while a Shorland armoured personnel carrier can be seen at the far end of the road. Courtesy: *The Sunday Island.*

2. BTR 152 command vehicle in Jaffna during the height of the militants' 'offensive' in 1984. Courtesy: *Sunday Island*

3. Sri Lanka Air Force helicopters. Bell 206 in the foreground followed by Bell 212 and Bell 412. Courtesy: *Sunday Island*

4. Cordon and Search Jaffna 1984. Courtesy: *Sunday Island*

5. Unexploded bomb weighing 55kg found in Jaffna after one of the earliest bombing raids on the city. Courtesy: *Saturday Review.*

6. Troops assembled in Palali before 'Operation Liberation. In the foreground is a Saracen APC and behind it a Saladin armoured car. In the far distance is a Buffel APC. Photo credit: Raj Vijayasiri

7. Weapons captured from the LTTE during 'Operation Liberation'. Photo credit: Raj Vijayasiri

8. Wounded soldier. Photo credit: Raj Vijayasiri

9. Special Forces establishing a blocking position south of Point Pedro during 'Operation Liberation'. Photo Credit: Raj Vijayasiri

10. Sunil, Nimal and Priyantha, the three Sri Lankan soldiers captured by the LTTE during the assault on the

telecommunications centre in June 1987. Courtesy: *Saturday Review*

11. Barrel bombs ready to be loaded. Note the wooden fins and the wheels. The former provided a measure of stability in the air and the latter enabled the bombs to be loaded into and pushed out of an aircraft with ease. Photo credit: Dr. Gamini Goonetilleke.

12. Sri Lankan troops alighting from a transport plane. These were the workhorses of the Air Force, transporting troops and supplies to the beleaguered northern garrisons. They were also used to drop barrel bombs. Photo Credit: Raj Vijayasiri.

13. Sri Lankan Special Forces on their way to a mission, riding a Saracen armoured personnel Carrier. Photo credit: Raj Vijayasiri.

14. General Kobbekaduwa with Major Gamini Hettiarachchi somewhere to the south of Point Pedro during Operation Liberation. Photo credit; Raj Vijayasiri.

Image no. 1

Image no. 2

Image no. 3.

Image no. 4.

Image no. 5

Image no. 6

Image no. 7

Image no. 8.

Image no. 9.

Image no. 10.

Image no. 11.

Image no. 12

Image no. 13.

Image no. 14

Bibliography

Books, Pamphlets and Reports

Ainley, Henry, *In Order to Die*, London: Burke Publishing, 1955.

Alles, A. C., *The J.V.P. 1969-1989*, Colombo: Lake House, 1990.

Baldaeus, Phillipus, *A True and Exact Description of the Great Island of Ceylon*, a New and Unabridged Translation from the edition of 1672, published as *Ceylon Historical Journal* vol. 8, July 1959 - April 1959.

Balasingham, Anton, *War and Peace: Armed Struggle and Peace Efforts of Liberation Tigers*, Mitcham: Fairmax Publishing Limited, 2004.

Balasingham, Adele, *Women Fighters of Liberation Tigers*, Jaffna : Thasan Printers, 1993.

Blodgett, Brian, *Sri Lanka's Military: the Search for a Mission*, San Diego, California: Aventine Press , 2004.

Chandraprema, C. A., *Gota's War: the Crushing of Tamil Tiger Terrorism in Sri Lanka*, Colombo: Ranjan Wijeratne Foundation, 2012.

Cordiner, James, *A Description of Ceylon, an Account of the Country, Inhabitants and Natural Productions*, New Delhi: Navrang 1983, (First Published by Longman, Hurst, Rees and Orme, Aberdeen, 1807) v.1.

De Silva, Gerry, *A Most Noble Profession*, Colombo: International Book House, 2011.

Dissanayake, T.D.S.A., *War or Peace in Sri Lanka*, Colombo: Popular Prakashan, 1995, vol. 2.

Dixit, J. N., Assignment *Colombo*, Colombo: Vijitha Yapa, 1998.

Emergency '79, Pamphlet published by the Movement for Inter Religious Justice and Equality, Kandy, 1980.

Ghosh, P.A., *Ethnic Conflict in Sri Lanka and Role of the Indian Peace Keeping Force*,
http://books.google.com.au/books?id=YZscr75ijq8C&pg=PA131&lpg=PA1
31&dq=operation+checkmate+IPKF&source=bl&ots=0XZJyh-
Run&sig=LSoPlhN80q_ayOqT_JmQF2lFVnc&hl=en&sa=X&ei=Ds-
pT4S4H8uuiQfcvuWtAw&ved=0CFUQ6AEwAw#v=onepage&q=operatio
n%20checkmate%20IPKF&f=false

Goonetilleke, Gamini, *In the Line of Duty; Life and Times of a Surgeon in War and Peace*, Colombo: Unigraphics, 2008.

Guneratne, Malinga H., *For a Sovereign State*, Ratmalana, Sri Lanka: Sarvodaya Book Publishing Services, 1988.

Gunaratna, Rohan, *War and Peace in Sri Lanka*, Colombo: Institute of Fundamental Studies, 1987.

Gunaratna, Rohan, *Indian Intervention in Sri Lanka: the Role of India's Intelligence* Agencies Colombo: South Asian Network on Conflict Research, 1993.

Guneratne, Major-General Kamal, *The Road to Nandikadal: The True Story of Defeating Tamil Tigers*, Colombo: Vijitha Yapa, 2016.

Hoole, Rajan (et al), *Broken Palmyra: The Tamil Crisis in Sri Lanka, an Inside Account*, Ratmalana, Sri Lanka: Sri Lanka Studies Institute, 1992.

Horowitz, Donald L., *Coup Theories and Officers' Motives: Sri Lanka in Comparative Perspective*, Princeton, NJ: Princeton University Press, 1980.

Jayatunge, Ruwan, *Sangramayen Pasu: Ealam Yuddhayata Muhuna Dun Soldaduwange Anuwedaneeya Katha*, Colombo: Sarasavi, 2007.

Leary, Virginia A., *Ethnic Conflict and Violence in Sri Lanka*, report of a mission to Sri Lanka n July-August 1981 on behalf of the International Commission of Jurists, International Commission of Jurists, 1983.

Little, David, *The Invention of Enmity*, Washington D.C.: U.S. Institute of Peace, 1994.

Mendis, L. M. H., *Assignment Peace, in the name of the Motherland*, Nugegoda, Sri Lanka: Author publication, 2009.

Miller, Phil, *Britain's Dirty War against the Tamil People 1979 -2009*, Bremen: International Human Rights Association, 2015.

Munasinghe, Rohitha, *Eliyakanda Wada Kandawura*, Colombo : Godage Brothers, 2000.

Munasinghe, Sarath, *A soldier's Version*, Colombo: Author publication, 2000.

Muthukumaru, Anton, *The Military History of Ceylon – an Outline*, Delhi: Navrang, 1987.

Narayan Swamy, M. R., *Inside an Elusive Mind: Prabhakaran*, Delhi: Konark Publishers, 2003.

--------------------- *Tigers of Lanka: from Boys to Guerrillas*, Delhi: Konark Publishers, 1994.

O'Ballance, Edgar, *Cyanide War: the Tamil Insurrection in Sri Lanka* 1973-88, London: Brassey's, 1989.

Ponniah, S., Satyagraha: *The Freedom Movement of the Tamils in Ceylon*, Jaffna: A. Kandiah, 1963.

Piyadasa, L., *Sri Lanka: the Holocaust and After,* London: Marram Books, 1984.

Ranatunge, Cyril, *Adventurous Journey*, Colombo: Vijitha Yapa, 2009.

Senanayake, Thilak, *Vadamarachchi Vimukthi Meheyuma Saha Uthure Satan*, Colombo: Godage Publishers, 2004.

Senaratne, Jagath P., *Sri Lanka Air Force: a Historical Retrospect*, 1985-1987, Colombo: Sri Lanka Air Force, 1998, volume 2.

----------------------- *Sri Lanka Armoured Corps: 60 Years of History* 1955 - 2015, Sri Lanka Armoured Corps, 2015

Seneviratne, Colonel Susantha and Harischandra, Ruwan, *Negenahira Mudhagath Meheyuma*, Colombo: Pahana Publishers, 2003.

Singh, Depinder, *The IPKF in Sri Lanka*, Delhi: Trishul Publications: Delhi, 2002.

Sivanayagam, S., *Sri Lanka: Witness to History, a Journalist's Memoirs 1930 - 2004*, London: Sivayogam, 2005.

Smith, Tim, *Reluctant Mercenary: the Reflections of an Ex-Army Helicopter Pilot in the Anti-Terrorist War in Sri Lanka*, Sussex: The Book Guild Ltd., 2002.

Sri Lanka Army 50 years on, Colombo: Sri Lanka Army, 1999.

Sri Lanka's Eastern Province: Land, Development, Conflict, Crisis Group Asia Report No. 159, 15 October 2008, Appendix C p.36

The Sri Lanka Navy: a Pictorial History of the Navy in Sri Lanka 1937-1998, Sri Lanka Navy, 1998.

Vittachi, Tarzie, *Emergency '58: The Story of the Ceylon Race Riots*, London: Andre Deutsch, London, 1958.

Wilson, A .J., *Sri Lankan Tamil Nationalism: Its Origins and Development in the 19th and 20th Centuries, London*: C. Hurst and Co., 2000.

Wilson, Peter, *Long Drive through a short War: reporting on the Iraq War*, South Yarra, Victoria: Hardie Grant Books, 2004.

Articles and Websites

'1st Reconnaissance Regiment Sri Lanka Armoured Corps', *Army Magazine*, 30th Anniversary Issue, October 10, 1979, p.23.

'50 killed in Tamil Raids', *The Sun*, 3. 6. 85.

'59 Tamils murdered', *Sun* , 21. 5. 85.

'80 Tamils die in Lanka Raids', *Herald*, 6. 6. 85.

'84 Reported slain as guerrillas raid Sri Lanka Farms', *Toronto Star*, 1. 12. 84.

Abeypala, H. W., 'Black Saturday's Slaughter House', *Weekend*, 15. 2. 87, p. 6 and p. 11.

Abraham, Thomas, 'Pounding Jaffna', *Frontline*, 2-15.3.91, pp.56-57.

'Adampan Hospital bombed,' *Saturday Review*, 14. 2. 87, p. 8.

Amato, Edward, 'Tail of the Dragon: The Sri Lankan Efforts to Subdue the Liberation Tigers of Tamil Ealam', Master's Thesis, US Army Command and General Staff College, Fort Leavenworth, Kansas, 2002.

'A new Spiral of Violence', *Asiaweek*, 24. 5. 85.

'Army Officer for Jaffna', *Malaysian Straits Times*, 29. 2. 84.

'Army 'Guilty' in Sri Lanka', *Herald*, 16. 8. 84.

Athas, Iqbal, 'Sri Lanka Strengthens Defence Forces', *Jane's Defence Weekly*, 3, 2, (12. 1. 85) p.45.

------------------ 'Operation Giant Step', *Weekend*, 22. 2. 87, p.7.

----------------- 'My Longest Day', *Weekend*, 11. 5. 86, p. 6.

----------------- 'The Fear of living dangerously', *Weekend*, 3. 8. 86, p. 11 and p. 23.

----------------- 'How Forces React to the Terrorists' Grand Design', *Weekend*, 28. 9. 86, p. 10 and p. 18.

---------- 'Rendezvous in Madras to Receive seized Arms', *Weekend*, 8. 3. 87, p. 7 & 21.

---------- 'The Stinging Bee Experience', *Weekend*, 14. 9. 86, p. 6 and 10.

---------- 'Peace through the Ballot?', *Weekend*, 15. 3. 87, p. 7.

---------- 'Bid to Internationalise Hostage Crisis by Tigers', *Weekend*, 29. 3. 87, p. 7 and p. 23.

---------- 'The Vadamarachchi Landing', *Weekend*, 31. 5. 87, p. 6.

---------- 'The Post Vadamarachchi Crisis', *Weekend*, 14. 6. 87, p. 6.

---------- 'How Operation Jellyfish Stung India's Flotilla at the Kutch', *Weekend*, 7. 6. 87, p. 6 and p. 23.

---------- 'Tigers Explode Peace at Nelliady', *Weekend*, 12. 7. 87, p. 6 and p. 11.

---------- 'The "Day After" Mood Plagues Jaffna', *Weekend*, 19. 7. 87, p. 6 and p. 19.

---------- 'Island Pride and Prejudice', *Weekend*, 24. 7. 87, p. 6.

'Bearing the Blockade', *India Today*, 15. 2. 87, pp. 24-27.

'Bleeding Statistics', *Saturday Review*, 4. 1. 86, p. 4.

'Blood, More Blood, on our Hands', *Saturday Review*, 18. 5. 85, p. 1.

Bob, Dilip, 'A Bloodied Accord', *India Today*, 15. 11. 87, http://indiatoday.intoday.in/story/after-16-days-of-bloody-battle-ipkf-finally-captures-ltte-stronghold-jaffna/1/337703.html (Accessed 1. 8. 16) Brown, Derek, 'Jaffna Reality – Two Strange Forms', *The Guardian*, 15. 6. 87, reproduced in the *Saturday Review*, 11. 7. 87, p. 4.

'Cannon Fodder', *Saturday Review*, 6. 6. 87, p. 1.

'Civilians? Hang them!', *Saturday Review*, 23. 5. 87, p. 8.

Clairborne,William, 'Tamils Hide in Fear as Troops take Revenge', *The Age*, 15. 8. 84.

Colonisation and Demographic Changes in the Trincomalee District and its Effects on the Tamil Speaking People,' University Teachers for Human Rights (Jaffna), Report 11, Appendix 2, http://www.uthr.org/Reports/Report11/appendix2.htm

Cooper, Tom, 'Sri Lanka since 1971", http://www.acig.info/CMS/?option=com_content&task=view&id=174&Itemid=1 (Last accessed 23. 10 . 15).

Cruez, Dexter, 'Within the Jaws of the Tiger', *Weekend*, 26. 10 . 86, p. 8.

De Alwis, William and Abeywardene, Srimal, 'Eleven Killed by Bomb in CTO', *Ceylon Daily News*, 6. 8. 09, http://www.dailynews.lk/2009/06/08/fea10.asp (Last accessed 25.12.15).

'Death rains from the Skies', *Saturday Review*, 1. 3. 86, p. 1.

De Boer, Marno, 'Rhodesia's Approach to Counterinsurgency: a Preference for Killing', *Military Review*, November - December 2011, pp. 35 - 45.

De Silva, Mervyn and Venkatramani, S., 'Reign of Terror', *India Today*, 31. 12. 84, pp. 22-23.

De Silva, Mervyn, 'Operation Turnaround or Turnabout?', *Lanka Guardian* 1. 6. 86, p. 3 & p. 6.

'Don Mithuna', 'It will Only be a War of Attrition if They Seek a Military Way Out', *Weekend*, 17. 11. 85, p. 6.

'Economic Sanctions Against Jaffna', *Saturday Review*, 10. 1. 87, p. 12.

Edirisinghe, Roland, 'Sri Lanka Links Rebels' Base to India', *Sydney Morning Herald*, 12. 1. 85.

Elliot, John, 'India Warns Sri Lanka of Offensive', Lanka *Guardian*, 1. 12. 86, p. 3.

---------------- 'Battle for Tamil Hearts, Minds and Stomachs', *Financial Times*, 6. 6. 87, reproduced in *Lanka Guardian*, 15. 6. 87, p. 8 and p. 11.

Ferdinando, Shamindra, 'Armless Veteran speaks of War and Peace', *The Island*, 27. 10. 2002, http://www.island.lk/2002/10/27/featur08.html (Last accessed on 10. 5. 16)

-------------------------------- 'Tragedy in the Wake of Victory at Velvetithurai', *The Island*, 8. 1. 13, http://www.island.lk/index.php?page_cat=article-details&page=article-details&code_title=69950 (Last accessed 10. 4. 13).

-----------------------------, 'Sri Lanka: the War on Terror Revisited; Sri Lankans Conduct Live Firing Exercises ahead of operation Liberation', *The Island*, 20. 6. 13, http://slwaronterror.blogspot.com.au/2013/06/israelis-conduct-live-firing-exercises.html (Last accessed 15. 12. 15)

---------------------------------- 'Army Loses Hero of Kokilai Battle', *The Island*, 10. 7. 13, http://pdfs.island.lk/defence/20130710_153.html (Last accessed 24. 4. 17)

Fernando, Austin, 'Revisiting Aranthalawa Bhikku Massacre, *Colombo Telegraph*, 2. 6. 2012, https://www.colombotelegraph.com/index.php/revisiting-aranthalawa-bhikku-massacre/ (Last accessed 25. 3. 16)

Fishlock, Trevor, 'Economy drained as Tamils flee city', *Australian*, 4. 1. 85.

'From Manal Aru to Weli Oya and the Spirit of July 1983', University Teachers for Human Rights (Jaffna), Special Report 5, http://www.uthr.org/SpecialReports/spreport5.htm#_Toc51256 9422 (Last accessed 5. 12. 15)

'Grain Cannot be Moved into Jaffna', *Lanka Guardian*, 1. 2. 1987, p. 12.

Graves, David, 'Troops, tackling rebels in divided Sri Lanka terrorise Tamils', *Daily Telegraph*, 17. 12. 84

Gupta, Shekar, 'Terror Tactics', *India Today*, 15. 10. 85, pp. 50-56.

----------------- 'Haven in India for Lankan Guerrillas', *Sydney Morning Herald*, 14. 4. 84.

Hamlyn, Michael, ''The Boys' Keep Army Behind Barricades in Fight for Tamil State', *The Australian,* 22. 3. 86.

Hamlyn, Michael, 'Tiger Guerrillas Step into Rulers Role: Jaffna Tamils prepare for post-settlement role in Sri Lanka', *The Times*, 18. 9. 86.

------------------------- 'Tamils step up attacks on civilians: the Communal Conflict in Sri Lanka', *The Times*, 11. 3. 86.

------------------------- 'Tamil Attacks Harden Attitudes: Sri Lankan Despair over Ethnic Crisis', *The Times*, 22. 4. 87.

------------------------ 'Slow Struggle on Jaffna Peninsula', *The Times*, 1. 6. 87.

Hanif, Jehan, 'Massacre in Jaffna: TELO Man Tells All', *The Island*, 14. 9. 86, p. 13.

Hawksley, Humphrey, 'How the Battle for Trinco was Won', *Lanka Guardian*, 1. 7. 1986, p. 13 & p. 24.

Humphrey Hawkesley, 'Tamil Guerrillas Kill 7 in Troop Convoy Ambush', *The Age*, 3. 4. 86.

---------------------------- "Sri Lanka Resumes Air Raids on Tamil Villages", *The Age*, 29. 3. 86.

---------------------------- "Tigers find teeth for a war at sea", *The Age*, 5. 3. 86.

Ibrahim, Hana, 'Fresh Attacks on Police Stations in North Repulsed', *Daily Mirror*, 15. 8. 84.

'India Cracks the Whip', *Asiaweek*, 23. 11. 86, p. 33.

'Inside a TELO training Camp', *Weekend*, 11. 11. 84, p. 1 and p. 5.

'Inside Jaffna Fort', *Saturday Review*, 4. 10. 86, p. 11.

'Inside the Jaffna Fort: Battle of Nerves', *Ceylon Daily News*, 11. 9. 86, reproduced in *Saturday Review*, 20. 9. 86, p. 6.

Ismail, Qadri, 'Military Option and its Aftermath', *Sunday Times*, 7. 6. 87, p. 5.

'Israeli, British Agents Helping Lankan Forces', *New Sunday Times*, 12. 8. 84.

'Is Rajiv Gandhi Being Misled?', *The Island*, 5. 10. 86, p .9.

'Is there a Way Out?', *Asiaweek*, 21. 6. 85, pp. 32-39.

Jayaweera, Neville, 'In to the Turbulence of Jaffna: a Chapter Extracted from the Author's Unpublished Memoirs titled 'Dilemmas'', *The Island*, http://www.island.lk/2008/10/05/features2.html (Last accessed 25. 12. 15).

'Jaffna Bombed', *Saturday Review*', 22. 2. 86, p. 1.

'Jaffna under Siege', *Saturday Review*, 25. 1. 86, p. 1.

Jeyaraj, D. B. S, 'LTTE Ascendant in the East', *The Island*, 9 .3. 97, p. 13.

---------------------- 'Ten Years Ago: The Kaithady Explosion', *The Island*, 16. 2. 97, p. 16.

-------------------- 'Birth and Growth of the Black Tiger Suicide Squad', *The Island*, 13. 7. 97, p. 16.

Johnson, Robert Craig, 'Tigers and Lions in Paradise: the Enduring Agony of the Civil War n Sri Lanka'", http://worldatwar.net/chandelle/v3/v3n3/articles/srilanka.html (Last accessed 23. 10. 15).

Joshi, Manoj, 'A Base for all Seasons: how LTTE used Tamil Nadu', *Frontline*, 3-16. 8. 91, pp. 21-23.

'J. R. breaks his Silence: I feared a Military Coup in'87', Interview with Vijitha Yapa, *Sunday Times*, 11. 2. 90, pp. 14-15.

Karniol, Robert, 'Rocket Boost for Sri Lanka', *Jane's Defence Weekly*, 28. 6. 2000.

'Kilinochchi Faces Famine', *Saturday Review*, 7. 3. 87, p. 8.

'Killed in Action', *Saturday Review*, 23. 5. 87, p. 1.

Kulatunga, Aruna and Wijeratne, Premalal, 'The Train Tragedy", *Sun*, 22. 1. 85

Kulkarni, V. G., 'The Military Modernises to Meet Rebel Threat', *Far Eastern Economic Review*, 12. 6. 86, pp. 29-31.

Lalith Athulathmudali's Interview with Qadri Ismail, *The Island*, 16. 6. 85, p. 9.

'Latest anti-Tamil violence claims 56', *The Australian*, 20. 5. 85.

'Leading Tiger Killed in Adampan', *The Island International*, 22. 10. 86, p. 1.

McDonald, Robert, 'Eye Witness in Jaffna', *Pacific Defense Reporter*, August 1987, p. 27.

'Manampitiya Police Guard Room Attacked-2 PCs killed', *The Island*, 21. 5. 85.

Mannar Tragedy", Saturday Review, 22. 12. 84, p. 3.

Mannar Battle Victims buried', *The Island International*, 22. 10. 86, p. 1.

Manor, James and Segal, Gerald, 'Causes of Conflict: Sri Lanka and Indian Ocean Strategy', *Asian Survey*, 25, no. 12 (December 1985), pp. 1165-1185.

Marks, Thomas A., "Insurgency and Counterinsurgency", *Issues and Studies*, August 1986, pp. 63-102.

----------------------- 'Sri Lanka's Special Forces', *Soldier of Fortune*, July 1988, pp. 32-9.

----------------------- 'Counter Insurgency in Sri Lanka, Asia's Dirty Little War', *Soldier of Fortune*, Feb. 1987, pp. 38-47.

'Military Problem', Saturday Review, 1. 2. 86, p. 1.

'Military Training in Tamil Nadu and India', *The Island*, 5. 10. 86, p. 9.

'Monitoring Committee on Batticaloa', *Saturday Review*, 18. 1. 86, p. 5.

'More than 50 Killed in Tamil Violence', *Canberra Times*, 20. 4. 84.

Naqvi, Sayeed, 'How the BBC Man Faced Jayawardena's Cannon', in S. Sivanayagam (ed) *Tamil Information* (published for Private circulation), Madras, 1. 9. 84, vol. 1, nos. 4 and 5, p. 17 (The article originally appeared in the *Indian Express*). pp. 16 - 17.

'Nine Die in Sri Lanka Riot', *The Australian*, 12. 4. 84.

'"No Community will Benefit from Provincial Councils" - Sirima', *Lanka Guardian*, 15. 8. 86, p. 8.

'Now, the Brigadier Speaks', *Saturday Review*, 25. 8. 84, p.12.

'On the Northern Front (Reuters Despatches)', *Lanka Guardian*, 1. 6. 1986, p. 4.

'Operation Blue Star' *Saturday Review*, 30. 5. 87, p. 1.

'Operation Liberation Encircles Key LTTE Post', *Daily News*, 28. 5. 87, p. 1.

'Operation Search and Destroy Launched', *Weekend*, 9. 12. 84.

'Operation Tiger: the Reasons Behind', *The Island* 7. 12. 86, p. 9 (courtesy *Amrita Bazaar Patrika*).

Palihawadana, Norman, 'Terrorists Get Away with Jeeps From Mahaveli Maduru Oya Schemes', *Island* 19.7.85.

Parthasarathy, Malini, 'A Military and Political Misadventure', *Frontline*, May 31 – June 13 1986. p. 19.

Peiris, Denzil, 'Colombo Rides the Tiger', *Weekend*, 3. 3. 85, p. 6 and 22.

Perera, Elmo, 'Bloodlust of the brutal Tigers', *Weekend*, 20. 03 1987, P. 8 and p. 21.

R. C., 'All Quiet on the Northern Front', *Saturday Review*, 30. 5. 87, p. 4

Rao, P. Venkateshwar, 'Ethnic Conflict in Sri Lanka: India's Role and Perception', Asian Survey, Vol. 28, No. 4 (Apr., 1988), pp. 419-36.

Richardson, Michael, 'Sri Lanka Accuses India of 'invasion'', *Age*, 31. 1. 85.

Rodrigo, Malaka, 'Kandula - the Little Elephant of the Army'', *Sunday Times*, 21. 06. 09, http://www.sundaytimes.lk/090621/Plus/sundaytimesplus_13.h tml (Last accessed 16. 7. 16)

Samath, Faizal, 'Terrorist Training Camp at Jaffna', *The Island*, 14. 7. 85, p. 1 & 2.

'Security Forces Secure Large Areas', *Sunday Observer*, 15. 2. 87, p. 1

'Security Operations Come to a Halt', *The Island*, 21. 5. 86, p. 1.

'See You in Jaffna', *Saturday Review*, 21. 2. 87, p. 8.

Silver, Eric, "Tamils Main Losers in Unwinnable War", *Canberra Times*, 18. 8. 84.

'Six Servicemen die in ambush", *Daily Mirror*, 13. 8. 84.

'Siva Tells the Hindu – Jaffna Area a Prison', *The Sun*, 21. 1. 85.

'SLAF Chopper Comes Under Sniper Fire', *Weekend*, 25. 11. 84, p. 1.

Sri Lanka (Ceylon) News-Letter published by the High Commission of the Democratic Socialist Republic of Sri Lanka in Canberra, 10th December 1984, P. 10

'Sri Lanka 'Forced' to Seek Israeli Help', *The Straits Times*, 3. 7. 84.

'Sri Lankan Guerrillas Raid Two More Banks', *Singapore Straits Times*, 10. 8. 84.

'Sri Lanka: the Siege Within', *India Today*, June 15, 1989, p. 121

"Sri Lankan Separatists Launch Attacks", *New Straits Times*, 7. 8. 84.

Sri Lanka Situation Report, Tamil information and Research Unit, Madras, India, Issue nos. 9, (15. 4. 86), 11 (15. 5. 86), 13 (15. 6. 86), 14 (1. 7. 86), 15 (15. 7. 86) and 16 (1. 8. 86).

Steinemann, Peter, 'The Sri Lanka Air Force', *Asian Defence Journal*, Feb. 1993, pp. 52-61.

Suguro, Suvendrinie, 'More Army Camps Planned', Daily *Observer*, 31. 12. 84.

Swain, Jon, 'Face to Face with the Guerrilla Commander: Cyanide Martyrs Bar Way to Peace', *Sunday Times*, 10. 8. 86, reproduced in *Lanka Guardian*, 1. 9. 86, pp. 11-12.

-------------- 'Children suffer horrific burns in Army's Offensive Against Tamil Guerrillas: Sri Lanka Ends Fighting to Give Peace a Chance', *Sunday Times*, 14. 6. 87.

'Tactical Offensive on the Diplomatic Front or Tactical Retreat on the Domestic?', *Lanka Guardian*, 15. 10. 86, p. 3.

'Taraki' (D. Sivaram), "The Cat, a Bell and a Few Strategists", *Sunday Times*, 20. 4. 97, p. 7.

'Tamil Guerrillas Kill 57 villagers', *Canberra Times*, 3. 12. 84.

'Tamils Flee as Villages are Burnt and Looted', *The Age*, 12. 6. 85.

'Tamils Hit at village', *The Age*, 13. 6. 85.

Tamil Rebels Strike Convoy', *The Australian*, 6. 12. 84.

'Tamil Terrorists Kill 150, Wound 300 in Sacred City Attack', *Australian* 15. 5 .85.

'Tamil Killings a Reprisal for Earlier Village Deaths', *Australian*, 15. 5. 85.

'Telecom soldiers charred', *Saturday Review*, 6. 6. 87, p. 8.

'Terrorists planning Easter Invasion', *Daily News*, 28. 2. 85.

'The Army will stay on', (Lalith Athulathmudali's interview with *India Today*'s Dilip Bob and S. V. Venkatramani), *Lanka Guardian*, 1. 7. 87, pp. 9-10.

'The Day of the lions (Jackals?)', *Saturday Review*, 6. 6. 87, p.2

'The Generals Lay a Trap', *Asiaweek*, 14. 6. 7, p. 21.

'The Know-how to Combat Terrorism', *Sun*, 26. 1. 85.

'The Murder of Alfred Duraippah', University Teachers for Human Rights (Jaffna), http://www.uthr.org/Book/CHA02.htm#_Toc527947383 (Last accessed 14. 4. 12).

'The Pirapaharan Phenomenon', chapter 15.
http://www.sangam.org/articles/view2/?uid=539 (Last accessed 21.12.15)

'The Pirapaharan Phenomenon', Chapter 16,
http://www.sangam.org/articles/view/?id=45 (accessed 15. 5. 12)

'Pirapaharan Phenomenon,' chapter 23,
http://www.sangam.org/articles/view2/?uid=633 (Last accessed 12.12.15)

'The Pirapaharan Phenomenon', Chapter 25,
http://www.sangam.org/articles/view2/?uid=645 (Last accessed 26.12.15).

'The Pirapaharan Phenomenon', Chapter 30,
http://www.sangam.org/articles/view/?id=212 (Last accessed 23. 11. 12)

'They Must be Defeated! Jayawardena Lambasts Tamil Rebels and India', *Time*, 11. 5. 87, p. 14.

'Tigers at Bay', *The Economist*, 28. 2. 87, p. 28.

'Tigers Hold Troops as Captive Force', Reuters report, *Weekend*, 21. 6. 87, p. 11.

'Tigers and Lions in Paradise: the Enduring Agony of the Civil War in Sri Lanka',
http://worldatwar.net/chandelle/v3/v3n3/articles/srilanka.html accessed on 20. 12. 15.

'Tigers V-sign', *Saturday Review*, 20. 4. 85, p. 8.

'Thousands Answer Army's call', *Weekend*, 9. 12. 84, p. 1.

'Top Secret Camp for Terrorists', *The Island*, 5. 10. 86, p. 9 and p. 15.

'Truly a Hero of Our Times', http://www.slnewsonline.net/kobbe.htm (Last accessed 14. 12. 15).

'Vadamarachchi Operation: The Missing Generation', *Saturday Review*, 20. 6. 87, p. 3, P. 4, & p. 9.

'Velvetithurai, A Fishing Village Victim of Pogrom', *Financial Times*, 21. 8. 84, reproduced in S. Sivanayagam (ed), 'Tamil Information', (published for private circulation) Madras, 1. 9. 84, vol. 1. Nos. 4 and 5, pp. 12 - 13.

Venkatramani, S. H., 'Battle lines', *India Today*, 15. 12. 84.

------------------------ 'Taming the Tigers', *India Today*, 30. 11. 1986, pp. 22-3.

-------------------------- 'Bearing the Blockade', *India Today*, 15. 2. 87, pp. 25 - 28.

Venkatnarayan, S., 'A Visit to Jaffna', *The Island*, 15. 2. 87, p .6.

'Victory and Defeat', *Saturday Review*, 24. 6. 86, p. 2.

Vijayasiri, Raj, *A Critical analysis of the Sri Lankan Government's Counter Insurgency Campaign,* (Masters Thesis, Fort Leavenworth, Texas, 1990.

'Violence in Jaffna: Five killed', *New Straits Times*, 11. 4. 84.

'V.V.T. Massacre', *Saturday Review*, 25. 5. 85, pp. 5, 7 and 8.

'VVT the Facts', *Saturday Review*, 31. 5. 86, p. 8.

Weaver, Maryanne, 'Civil War Looms with Separatists', *The Australian*, 31. 1. 85.

Weisman, Steven, 'Terror on the Beach in Sri Lanka', *Sydney Morning Herald*, 12. 2. 85.

West, Julian, 'Passage to Jaffna', *Asiaweek*, 8. 3. 1991, pp. 13-22.

'We Will Defeat the Terrorists in One Year: JR tells BBC', *The Island*, 27. 10. 85, pp. 1 and 2.

Wimaladasa, Vilma, '86 Killed in Tamil Raid on Holy City', *Daily Telegraph*, 14. 5. 85.

Winchester, Simon, 'Behind the Lines With the Tamil Guerrillas', *The Australian*, 29. 6. 85.

'Yal-Devi blast: Death toll 39', *Sun*, 22. 1. 85.

Interviews

Brigadier Bahar Morseth, 21. 11. 15.

Colonel Kapila Ratnayake 13. 12. 15.

Lieutenant Colonel Modestus Fernando, 7. 8. 2016.

Lieutenant Sudesh Ranawaka 31. 1. 2016.

Index

www.ingramcontent.com/pod-product-compliance
Lightning Source LLC
Chambersburg PA
CBHW020833210326
41598CB00019B/1887